YOUR BRAIN IS (ALMOST) PERFECT

DR. READ MONTAGUE is a professor in the department of neuroscience at the Baylor College of Medicine, director of the Human Neuroimaging Lab, and director of the Center for Theoretical Neuroscience. He has been a fellow at the Institute for Advanced Study in Princeton, New Jersey. His research focuses on computational neuroscience—the connection between the physical mechanisms present in real neural tissue and the computational functions that these mechanisms embody.

Praise for *Your Brain Is (Almost) Perfect*

"A fascinating introduction to an important new area of research in the science of mind."

—Steven Pinker, Johnstone Family Professor, Psychology Department, Harvard University, and author of *The Blank Slate* and *How the Mind Works*

"Montague knows that cool reason is not enough to explain decisions: Emotions provide biological value and are an indispensable factor in how one chooses. Drawing on a vast amount of recent knowledge, some of which [is] coming from his own work, Montague makes a compelling case for the significance of that fact in everyday choices. Read this book and maybe you will discover why you buy what you buy."

—Antonio Damasio, Director, Brain and Creativity Institute, University of Southern California, and author of *Descartes' Error*

"Choosing is as basic a business as breathing. But how do brains make good choices and avoid stupid ones? A gifted neuroscientist, Read Montague, has spent the last decade plumbing these mysteries. Assembling a vast range of experimental results, he has now crafted a simple and coherent framework that demystifies choosing. Importantly, the approach hogties old philosophical dogmas, and invites us to look at ourselves anew, from the brain's point of view. The book is comprehensively informed, calmly unafraid, and heartwarmingly witty. It tells a gripping story about what makes me, me."

—Patricia Churchland, Chair, Philosophy Department,
University of California, San Diego, and MacArthur "genius" award winner

"The question of how we make choices is one of the oldest in philosophy. Read Montague, a pioneer in this field, has for the first time tackled the problem from a neuroscientific perspective that not only makes it a valuable antidote to philosophy, but is fun to read as well."

—V. S. Ramachandran, M.D., Ph.D., Professor and Director,
University of California, San Diego, Center for Brain and Cognition, and
author of *The Man with the Phantom Twin*

"This book covers, and connects, insights from computer science, physics, evolutionary psychology and biology, psychology, economics, and of course neuroscience. In addition to the particular insights from the different fields, the connections Montague draws between them help us understand what the brain attempts to achieve and how it goes about doing so (taking us with him). The last part of the book describes in more detail a few amazing fMRI experiments, what was learned from them, and their larger implications . . . Defiantly a fascinating window into this new domain of science."

—Dan Ariely, Alfred P. Sloan Professor of Behavioral Economics,
Massachusetts Institute of Technology

"Montague deftly marries psychology and neuroscience . . . [and] adds new ideas to our understanding of how our brains compute."

—*Publishers Weekly*

"Provocative and accessible . . . the book spans several seldom-bridged worlds, from neuroscience to psychiatry, economics and social psychology, and does so with wit, precision, and elegance."

—*Nature*

Your Brain Is (Almost) Perfect

HOW WE MAKE DECISIONS

READ MONTAGUE

A PLUME BOOK

Previously published as *Why Choose This Book?*

PLUME
Published by Penguin Group
Penguin Group (USA) Inc., 375 Hudson Street, New York, New York 10014, U.S.A. • Penguin
Group (Canada), 90 Eglinton Avenue East, Suite 700, Toronto, Ontario, Canada M4P 2Y3 (a
division of Pearson Penguin Canada Inc.) • Penguin Books Ltd., 80 Strand, London WC2R 0RL,
England • Penguin Ireland, 25 St. Stephen's Green, Dublin 2, Ireland (a division of Penguin Books
Ltd.) • Penguin Group (Australia), 250 Camberwell Road, Camberwell, Victoria 3124, Australia
(a division of Pearson Australia Group Pty. Ltd.) • Penguin Books India Pvt. Ltd., 11 Community
Centre, Panchsheel Park, New Delhi – 110 017, India • Penguin Group (NZ), 67 Apollo Drive,
Rosedale, North Shore 0745, Auckland, New Zealand (a division of Pearson New
Zealand Ltd.) • Penguin Books (South Africa) (Pty.) Ltd., 24 Sturdee Avenue, Rosebank,
Johannesburg 2196, South Africa

Penguin Books Ltd., Registered Offices: 80 Strand, London WC2R 0RL, England

Published by Plume, a member of Penguin Group (USA) Inc. Previously published in a Dutton
edition under the title *Why Choose This Book?*

First Plume Printing, October 2007
10 9 8 7 6 5 4 3 2

 REGISTERED TRADEMARK—MARCA REGISTRADA

The Library of Congress has catalogued the Dutton edition as follows:

Montague, Read.
  Why choose this book? : how we make decisions / Read Montague.
     p. cm.
Includes bibliographical references and index.
ISBN 0-525-94982-8 (hc.)
ISBN 978-0-452-28884-3 (pbk.)
1. Choice (Psychology) 2. Decision making. I. Title.
BF611.M65 2006
153.8'3—dc22

                                                                  2006019015

Printed in the United States of America

Set in Minion
Original hardcover design by Katy Riegel

CONTENTS

WHY DID HE DO THAT?

WHY DID YOU CHOOSE this book? Maybe your choice felt driven by the cover design or a review or something more substantial like a past experience that changed your life. Perhaps the decision felt sort of random, without any clear intent one way or another. Nevertheless, you did choose, and choice means loss. For just a moment, you lost the chance to view other nearby books and instead opened this one. Your time and energy is limited and so your choice consumed precious resources. And as small as this particular investment may be, understanding your choice is important because your life is literally a compilation of billions of similar "choice moments" where one outcome is selected and others forgone. Who or what makes these choices? Why are some choices automatic and thoughtless while others feel like they are expressions of our free will? These questions strike at the very heart of what it means to be a thinking, feeling human, and yet our sheer familiarity with making choices hides their true nature from us. Choices are not magic, descending from some immaterial place like cost-free manna. Our brains and the experiences they harbor generate our choices, and this is why brain disease and injury can radically change the way we choose and consequently change who we are.

The sneaky secret about choice is that it's not about choice at all—it's about value. The act of choosing one action over another follows directly from the way the brain values both the external world and the internal world—our thoughts. And the surprise about valuation is that it arose because life runs on batteries and so decisions must be efficient. Value and efficiency. From these humble beginnings, life discovered subtle mechanisms of efficient valuation and embedded them in the human brain. It's these same mechanisms that constrain our choices, anchoring them firmly to our biological needs. Yet they also confer on us social agency—the freedom to choose, even to the point of denying our instincts for survival. What other species can and will die for an idea—even an idea read about in a book? So, good choice and read on.

WE WERE STUCK IN TRAFFIC on an unusually hot Thanksgiving Day in Atlanta. Our destination: the Georgia–Georgia Tech freshman football game, an annual gathering requiring the attendance of all school-age males in my family. The game is not particularly meaningful, since it's a freshmen-only game, but diehards love it because of the intrastate rivalry between the two schools. The traffic was locked in a column with no movement in sight. The car had no air conditioning. Sweat beads meandered down my forehead while the smells of the cracked vinyl bench seat lapped at my young nose. My hope of escaping car sickness drained rapidly until I noticed freedom just ahead—our cross street was about fifty yards away and totally absent of traffic. We needed only to make a right turn onto that street to escape the gridlock; the empty street beckoned. Apparently, my uncle also heard the call. He gripped the wheel, inhaled slowly, turned the wheels hard right, and drove up onto the sidewalk! Yikes! I gripped the seat and held my breath in exactly the same fashion as I did (do) just before the dentist's drill is brought to life.

"Never do what I do, boy," he said.

Which really meant, "I don't want to hear any commentary now or later."

The car bumped over the curb, maneuvered left around a nearby fire hydrant, and headed gingerly down the sidewalk toward the empty cross street. I sat in stunned silence as the car rolled forward and slowed, rolled forward and slowed. I felt a sense of relief each time my uncle braked before doorways that might empty an unsuspecting customer into our path. My initial shock morphed into a bizarre internal dialogue.

I could hear myself thinking: *Just look how carefully he drives along the sidewalk. He is being very safe around doorways.*

Safe around doorways! What was I thinking?

And my ridiculous conclusion followed: *And who can fault him for that?*

The reasonable answer to this question is almost anyone. Let's face it: Cars don't belong on sidewalks. However, I was locked into one of those childhood moments where things don't quite make sense and my mind strained to conjure some justification. Why was the football game so important? What if someone had stepped out in front of the car? Most pedestrians aren't exactly looking for a car on the sidewalk. Something was born in my mind that day about choices and the hidden values that must underlie them. How could my uncle risk possible arrest for "driving on sidewalk" to arrive at a freshman football game before halftime, and why did he value getting to the game more than following reasonable driving rules? I'm not sure what kind of punishment might accompany a conviction for "driving on sidewalk," but I don't think that that occurred to him at the time.

Which do you value more, your life or a shopping spree at the local grocery store? How about your daughter's life or half a glass of water on a hot day? These comparisons don't normally arise in everyday life, but the point is that our minds can value extremely different events, objects, actions, and intended actions against one another. Apples against oranges, you might say. Our minds are quite literally valuation machines. We can value separately the past, the present, and the future. We can even value things that did not happen—"If only I had chosen X . . ."—or have not happened yet—"Were I to choose X . . ."

The ability to evaluate different choices in the world emerged because it was an efficient solution to fundamental problems of survival—the most important problem being the need for flexibility in the face of uncertainty. The real world—the one with exposed roots, gnarly trees, unpredictable weather, and the vast ambiguities of most decisions—challenges every creature on our planet to find ways to be flexible. In this book, I detail mechanisms in our brains that guide our daily lives in this challenging world. I don't venture any deep arguments about their evolutionary origins other than my recurring focus on efficiency. My omission is meant not as a particular value judgment of what has come to be called evolutionary psychology; that approach has produced valuable insights into the constraints that shaped our cognition, but identifying evolutionary constraints is just one aspect of understanding how we think. Instead, my emphasis is on the here and now: those brain mechanisms and the mental software they implement that guide individual choices moment to moment. Efficiency is not the only important feature of life that shaped our brain's computations, but as a single idea it provides one of the best ways to connect specific features of our brains' anatomy and physiology to our daily lives; it lets neuroscience reach out and touch some important parts of psychology. However, no one can understand anything important about biology without understanding Charles Darwin's idea of evolution by natural selection. As the late, great biologist Theodosius Dobzhansky wrote, "Nothing in biology makes sense except in the light of evolution."

THERE IS A MODERN approach to brain and mind that stands on the shoulders of evolutionary biology. It's called computational neuroscience, the field that studies the actual information processing supported by our brains. This field began embryonically in the mid-twentieth century, but has reemerged with gusto in recent years to address every part of our neural and mental function as information processing–computation. But computation without further insights is still not quite enough to account for how our brains generate the moment-to-moment operation of our minds. The idea of valuation is required: the ability to assign value and make choices based on those values. To connect computation to values to mental operations we care about, I propose a new guiding idea—efficient computation. Remarkably, the single property that all efficient computational systems require is goals, and computations with goals mean computations that can care about something. This is the most important feature that distinguishes biological computation from the work carried out by our personal computers—biological computations care. I don't mean mysterious caring without any physical basis, but valuing some computations more than others.

While even bacteria can value their near-term future and recent past, I will argue that humanity's special capacity to value arbitrary objects and behavioral acts confers on us a kind of behavioral superpower not rivaled in the nervous systems of any other species—we can choose to veto our instincts for survival based on an idea. We will explore this capacity, its consequences, and the neural mechanisms thought to underlie it. This extraordinary human quality has a grimmer side, and the computational neuroscience approach offers a new view of some awful diseases. Drug addiction, Parkinson's disease, and various obsessive thought disorders can be better understood from the perspective of computational neuroscience. In making these connections, it is my hope that clinicians and computer scientists will see common ground for new approaches to a range of mental and neurological problems. These same ideas will reveal computational roots underwriting our ability to trust, betray, and regret. Although I will show that we are fundamentally computational creatures, we aren't mindless automatons, but possess the capacity to choose.

I tip my hat to all the thousands of living scientists and thinkers who have contributed to the subjects that I touch upon (I'm counting on the nonliving ones not having access to e-mail to air their complaints about omissions). I have tried with the endnotes to give pointers straight into the literature for the intrepid reader, but the luxurious equivocation that all scientists use to guard their work is largely absent here.

As modern neuroscience continues to deliver on its promise of understanding the biology of the nervous system, it's very clear that neuroscience is not equipped on its own (or on its own terms) to comprehend fully how to connect its vast physical understanding of the brain to our mental software. Psychology, cognitive science, and formal theoretical work will always be needed for this job. Software problems are not always directly related to hardware problems— biological or otherwise. In many ways, psychology may have more to teach the physical scientists of the brain than the other way around. I suggest this position grudgingly, having started my life in science at the biophysical level, later perceiving the unavoidable importance of psychological-level descriptions of mental function and realizing quickly that I knew virtually nothing about real psychology. I have since tried to rectify the situation and hope this book will contribute a new perspective that appeals to people who have not previously considered the fundamental ways in which a mind can be understood as a computational device running on our brains.

YOUR BRAIN IS (ALMOST) PERFECT

I

COMPUTERS THAT CARE

How Desperation Built Value into Brains

Choose the best for the least.
Now, that's real economics.
<small>ANONYMOUS</small>

THE EXPLOSION HAD BARELY subsided when the cold Atlantic brine embraced the last signs of the great merchant vessel. Another loss added to the escalating toll being extracted from Allied shipping. Suddenly, radio traffic crackled as an encrypted message zipped away from the attacking submarine to the rest of the wolf pack; no shyness about talking here, prey was near and unable to escape. Cipher operators, sitting quietly in their U-boats and focusing intently on their strange keyboards, decoded the message as a call to kill, directing the other submarines where and when to strike. Another convoy was about to meet a terrible fate.

Two brains, one warm and one cold, steered the terrible effectiveness of U-boat attacks in the North Atlantic during World War II. One, warm to the touch and living, belonged to Admiral Karl Dönitz, chief architect of these assaults on Allied shipping, assaults that threatened to sever supply lines to England. The other, cold and unfeeling, belonged to Enigma, the device that encrypted and decrypted the radio messages, hiding their contents from all eavesdroppers not possessing its secrets. One had a goal—German victory over Britain—and the other computed blindly with no intent. But together they could not be ignored. For Britain, everything was at stake, so desperation and urgency were the motivators of the day. Torpedoes and shells were a problem, but one understood by all the naval commanders; the real threat was those messages. Those damnable computer-encrypted messages. At any cost, they must be cracked open and understood quickly. And this urgency was fueled further by the human need for revenge and justice. The Enigma machines didn't care about the drowning Allied seamen. Dönitz the warrior chose to kill, but he probably did care. His decision emerged from his brain—a special kind of computer, one that makes decisions in a way that can now be understood by computer science, although real brains are not much like the computers we have used since World War II.

TWO QUESTIONS FACE EVERY creature on our planet every moment of every day. "What is the value of my available choices?" And beyond that, "How much does each choice cost?" For a mobile creature, choosing is not optional. All outcomes, including doing nothing, are choices. At its core, moment-to-moment living is a problem of investment and returns, and we all want to choose so as to get the most return for the least investment. These problems seem straightforward until we consider the stakes involved for real creatures in the real world: life or death. This is why organisms are desperate. Failure means oblivion. Which creatures win in this world? Those that accurately estimate the costs and the long-term benefits of choices will be more efficient than those that don't—and in the long term these are the winners. This is why one sees valuation mechanisms present at every level in biological systems, from molecules all the way up to strategies for social exchange. What do valuation mechanisms give to a system? They let a system care. The system can care more about one event than another. And that's the key to building information-processing systems, biological or otherwise, that are truly adaptive—they must care about what they do.

At one time or another we all hate our computers. Well, we don't really hate our computers; we hate the programs they run, or more specifically their output. And the reason that we sometimes hate their output is that programs just don't care about it—they don't care about your intentions, and they don't really care about one answer any more than another answer. Just give them some input, hit a button, and presto—output. Quite efficient when things go right, yet the source of stomach acid, blood pressure spikes, and teeth-grinding when they don't. Who hasn't wiggled a mouse, let her finger slip off a button, or committed some similar computer sin only to find a document missing, the font Comic Sans infiltrating a business letter, or a spreadsheet appearing full of unrecognizable characters? And, at one time or another, computer programs have pushed us all to the edge.

The gripe here is that computer programs, exercising great power over our momentary moods, are sort of accidentally evil. Why? They have not been given the capacity to care—they don't have any goals. They have no capacity to know generally what we are trying to accomplish, and so when something unexpected, ambiguous, or just slightly odd arises they are stuck and we are frustrated. No one would rank this outcome as efficient, and it's not. Lots of wasted time and energy, which could have been saved if the programs were more flexible, if they "got the point" of what we intended to do and were guided by some overriding goal. If these programs had been creatures trying to survive, they would be dead because they don't know how to care about their behavior.

Modern computer programs do have design specifications, preassigned jobs that they are "programmed" to carry out. Some programs even express a modicum of flexibility in the way they work, forgiving or even anticipating typical mistakes like errant button presses. But goals are what they need. The reason is that a goal is not a fixed prescription that determines once and for all the way a system will behave. Instead, a goal is just like informed advice: It equips a system with a kind of guidance signal that can be queried in times of real need, times of ambiguity and uncertainty. And in those times, goals are essential to keep the system from flopping about, wasting all sorts of energy, just because it's a little bit uncertain about what to do next. So goals are efficient, and in the event that they are used for guidance, they define what the system cares about.

In this book, I will introduce principles of efficient computation and use them to connect neural function to psychological function. I will use these principles to understand how our minds are produced by a living computer made out of slimy parts that care and how this collection of parts generates the power to pursue goals, even goals inconsistent with its own survival. These principles also explain seemingly anomalous features of our brains: for example, why they are slow, imprecise, and noisy. Efficient computation also explains why our brains "should be" this way and why modern computers, whose ancestors include the World War II Enigma machines, get very hot and biological computers like our brains do not, yet perform feats of computation unmatched by any machine. But beyond this physical perspective, the principles of efficient computation provide another way to understand our minds as the cornerstone of the ultimate survival machine—but one possessing an ironic twist: the freedom to choose not to survive. In the service of an abstract idea, like a political goal, we can choose to kill ourselves. Yet, it's this same powerful capacity to override our instincts that infuses abstract ideas with meaning and gives them the ability to guide our choices. And all this is accomplished with a device that is merely warm—our brain. How can this be? What is the central difference between computers and brains?

THE HEADMASTER'S ASSESSMENT was as clear as it was harsh: *"If he is to stay at Public School, he must aim at becoming educated. If he is to be solely a Scientific Specialist, he is wasting his time at a Public School."* And exactly who was this erstwhile student? Alan Mathison Turing—mathematician, maverick, code-breaker during World War II, true godfather of computer science, and outstanding distance runner. Turing never tried to endear himself to people like his headmaster, and with this strength of mind he built the machines that decoded Enigma. Alan Turing should be considered the Darwin of the twentieth century. Darwin transformed a mystery—how life can acquire its diverse adaptations—into a problem. Turing also changed a mystery—how minds can arise from physical interactions—into a problem.

Turing graduated in mathematics from King's College at Cambridge University in 1934 and within a year had been elected a Fellow of King's College. But beyond this social honor, he also invented the cornerstone idea on which all modern computing machinery depends.

Today, the same basic idea also forms a central part of what is called the computational theory of mind. His now-classic paper "On Computable Numbers, with an Application to the Entscheidungsproblem (Halting problem)" was published in 1936 and presents his exceptionally clearheaded ideas about computation that form the basis of how we think about computation today. In that paper, Turing proposed and demonstrated that *any* step-by-step procedure (or *algorithm*) could be represented as a sequence of elementary computations. And I mean really elementary—like "Write down a 1," "Erase a 0," "Move three steps down the list of steps," and so on. Turing bundled these simple notions into the form of an imaginary device, an abstract machine now called a Turing machine, and used his new construction to show what it meant for a number to be computable. These ideas opened new areas in mathematics, but more importantly, the machine described by Turing, although operating only as a thought experiment on paper, was the modern idea of a computer.

Turing described clearly what it meant for a computing machine to compute. And his idea provided a new insight—there is a difference between the patterns of computations running on a device (e.g., software) and the device parts and their interactions that implement the computations (e.g., hardware). Turing's insight ultimately blurred the distinction between "physical device" and algorithm. What's really remarkable about his discovery is that, although it helped inspire the design and construction of modern electronic computers in the twentieth century, it is fundamentally a twenty-first-century idea. Despite his early death at the age of forty-one, Turing's time is still just beginning.

A Turing machine is a device that can simulate anything

computable, that is, anything that can be effectively described, including another Turing machine. An electronic computer supports the execution of word processing programs because it has another program, called the operating system, on which the word processing program runs. So the "parts" that support the word processing program as you sit and edit are themselves made out of logical operations implemented by the operating system. These in turn rest on other operations running on board the processor chip. Turing's idea resonates with the twenty-first-century Internet environment because programs can whiz around the world at the speed of light to download onto some device where they carry out their functions. There are now programs that simulate entire computer chips. An operating system like Windows XP can be "loaded on top of" this chip simulator, and application programs that normally run on top of Windows will run. Really quite amazing when you step back and let it sink in—I'm old enough to marvel at how far and fast computing has evolved since 1974, when I wrote my first programs. And here we have an example of program-simulated hardware (the chip) supporting an operating-system program that in turn supports the word processing program; a veritable program-fest. And these software devices, these machines, scoot along as patterns of ones and zeroes, all encoded as fluctuations in the electromagnetic spectrum. This world might have stunned Darwin, but Turing would have recognized his machines right away—and delighted in them. A description of an early presentation of his ideas shows this to be true. In describing the Turing machine concept he said: "[it] can be made to do the work of any special-purpose machine, that is to say to carry out any piece of computing, if a tape bearing suitable 'instructions' is inserted into it." He clearly saw that his proposal blurred the boundary between machine and algorithm, but it also clarified the difference between brain and mind.

The key step is Turing's idea to differentiate a step-by-step procedure from the underlying device that implements it, whether the device is made out of electronic parts or logical steps or Tinkertoys. Programs need the machines that support them, but are not equivalent to them. And it's this basic insight that Turing applied to the idea of what a mind "is." Turing conjectured that even our thoughts were equivalent to computational steps, only running on a very specific, biologically evolved device: our brains. This idea is now called the computational theory of mind (CTOM). It's easy to state but still profound: Your mind *is not* equal to your brain and the interaction of its parts, but your mind *is* equivalent to the information processing, the computations, supported by your brain. This is a beguiling notion; however, modern philosophers have charged that something essential is missing from this view: meaning. They say that simply equating mental phenomena with information processing leaves out the meaning that is so clearly associated with mental events like love, trust, desire, regret, and so on. They have a point—these mental events are indeed meaningful to all of us. We care about these things. We will rearrange our entire lives based on them. So the computational theory of mind is an outright failure, right? Wrong. It's just incomplete. The philosophers are right about one thing, the meaning part is missing. How can one computation carry meaning that is different from another computation's? In fact, how can a computation ever carry any meaning at all? Let's first get the computational-theory-of-mind idea straight.

In the West, we conceptualize our existence in terms of two basic, but separate, entities: body and mind. This idea, usually attributed to

René Descartes, is an old one and thought by many to be obvious. The "obviousness" of the idea arises because our perception presents to us our mental experience as though it emerges from nowhere. When I ask you the name of your maternal aunt, her name may "appear" to you, but where was it before I asked? And where does it go when you are no longer thinking about her or her name? Ideas seem to arise from nowhere and simply disappear when they are "out of mind." They are "here when they're here" and "gone when they're gone"—not very satisfying. Where is this place where all these thoughts are hiding? Thoughts have never seemed like real-world "stuff," such as dirt or food or sex or injuries. They just arise, sometimes cued by external events in the world and sometimes just popping in unannounced like pesky in-laws. Mindlike stuff just seems so different from stufflike stuff, and until the last century it has been difficult to question mind-stuff very clearly.

The movie *The Truman Show* captures this idea beautifully. In the movie, the character Truman (played by Jim Carrey) is born and raised in a completely controlled world. Unbeknownst to him and under the secret gaze of television cameras, his experiences are artificially manufactured. Worse, his life is being broadcast on worldwide television. The show is a megahit watched like some kind of über–soap opera by addicted followers of Truman's manufactured life. In an interview, Christof, the artiste-creator of the show, is asked why Truman, at the age of about thirty, is only then just becoming aware of the nature of the world around him. Christof's response is to the point: "We accept the reality of the world with which we are presented." Don't we all—at least until we stop taking our thoughts for granted. And so mind has always seemed like "something else."

For millennia, everyone thought that the mind did not exist in any physical form or place—it seemed so self-evident. Therefore, mental events like desires, plans, regrets, love, and even moral urges must reside "somewhere else." Hardly anyone questioned this view: Thoughts have no material basis; they live in some indescribable place or state of being and can never be captured by any physical description. This idea, while perhaps emotionally appealing, is incompatible with a literal mountain of facts about inheritance and the evolution of biological traits. So what do we do now? Throw up our hands and declare it all a great mystery and go home? Understanding the mind is far too important for such laziness. There are virtually no data to support such a radical disconnection between mind and body, a rupture that would separate the character of mental phenomena from the evolution of everything else that comprises us. Our elbows can evolve, but not our minds—sorry, just not feasible. In fact, if minds really could not be represented by physical interactions (and evolve like everything else), then there would be no point worrying about the operations of the brain, because the brain's mechanisms are physical and therefore describable and evolvable. If I thought that that thought was indescribable, this book would end here.

Still, something feels amiss. Operations of the mind such as trust, love, happiness, and their kin do feel abstract. So do a host of other thoughts and feelings. If mind and brain are intimately linked products of biological evolution, then how can the operation of the brain account for these mental experiences? This is where the computational theory of mind (CTOM) steps in to tell us that our mind is the pattern of information processing running a special kind of machine: our brain. And while CTOM does not account for all our experiences, it was the first workable idea about how to get mind from brain. Turing's idea ties patterns of computation (mindlike stuff) to physical interactions (stufflike stuff) taking place inside our skulls. Evolving different brains means evolving different mental capacities, so Turing's conjecture binds the biological evolution of the brain to the evolution of the brain's products, our mental events. Evolution selects for some psychological mechanism, and in doing so it also selects for a host of physical interactions in the brain. Mind and brain evolved hand in hand and continue to do so today. This is not a new idea; evolutionary biologists and some philosophers of mind have had this general perspective for many years. However, it was Turing's specific insight about computations and their distinction from the physical device that implements them that explained the difference between the mental and the physical.

The central idea is computation. All things "thoughtlike" are patterns of information stored, processed, and transformed by physical mechanisms in your brain. This is how something immaterial like an abstract thought can be grounded in the physical operation of the nervous system. Your mind is not equivalent to your brain; it is the result of information processing supported by your brain. But the relationship between mind and brain is not one-to-one. There are many levels of information processing in a living brain. In fact, much of our mental software is unperturbed by detailed changes in the underlying structure and function of the brain; a kind of fault tolerance is built into this robustness. Neurons can die, their connections retract, and many other physical changes in the brain can take place while our mental life runs along without a ripple in its performance. The fact that we replace almost every atom of our soft tissues every couple of years shows clearly that our minds and even our bodies, portrayed as patterns of information processing, are stable to enormous amounts of turnover of the underlying parts. We are like the ultimate eddy current, a flesh-and-bone holding pattern of body and thought, hovering for seventy to eighty years, and then dissolving back into the earth—pretty cool.

The idea of computation has now descended in its full glory on every bit of biology—biological mechanism is information processing. This idea began first in the mid-twentieth century in a famous paper by Max Delbrück and Linus Pauling. In that prescient paper, these two Nobel-laureates-to-be proposed that the key feature to biological reactions (we would now say biological information processing) was the interaction of complementary molecular structures—complementary like lock and key. This early proposal was resoundingly correct and represents the now-implicit view of all of modern molecular biology. Structure is function. The shape of a bioactive molecule defines the other molecules with which it may interact: those with complementary shapes. Shape defines function. The modern idea of computation adds an extra item. Now the idea reads "Structure is function is computation." So structure can now be seen as algorithm. Consequently, the evolution of biological structure is the evolution of algorithms. As simple as this shift in perspective may seem, its consequences run deep and will continue to emerge in the coming century. It's information processing all the way down and all the way up. From the way that antidepressant molecules like Prozac interact with receptors in the brain to the large-scale "mental effects" of this interaction as one's mood lifts—it can all be described as a vastly complex computation.

Think about it a bit and the implications keep rolling out. Biological systems possess layer upon layer of physical structure and hence layer upon layer of information-processing levels. Cells house DNA and DNA houses genes; genes build proteins; proteins build hearts, lungs, bones, and brains, which themselves operate to control bodily functions and moment-to-moment psychology. Patterns built on patterns built on patterns. Every bit of it is computation. Sound far-fetched? It's not really if you seek an example straight from the source—DNA.

Deoxyribonucleic acid (DNA) has intricate physical properties. It's a two-chain polymer made out of nitrogen-containing bases with an array of structural and chemical features. But the really important part about DNA is that it stores and processes information—meaning that on top of its myriad physical properties, DNA is a nanoscale computational device. And what's even more interesting is that DNA carries information for building little machines (proteins) that can translate the other information also encoded in its sequence. Instructions that know how to build machines that can read the instructions. Cute trick. All a cell needs to know is how to "jump-start" the process, and DNA can be used to build machines that read it; DNA is the ulti-mate "navel-gazer" molecule.

Humans don't engineer machines like this. It's like producing a set of instructions for building a car that contain instructions for building machines that read and translate the instructions for build-ing the car. If you gave me instructions for building a car (a truly fruitless act on your part), I would first need to learn how to read and interpret them. Through experience and feedback, I would build structure in my brain that allowed me to interpret and act upon the instructions. DNA is similar, except that it no longer has to be taught about decoding itself. Through eons of natural selection, it has already learned to interpret its own instructions and stored the answers within itself—embedded within it lives the "secret to its own decoding." I realize that's a cheesy line lifted from the movie *Contact*; however, it directs us to another high-altitude insight.

The insight is about life. It's the information processing that is equivalent to the moment-to-moment "living" that a cell does, not the parts on which that processing runs. This is simply Turing's idea applied to living cells. In my opinion, the early and confusing ideas about vitalism—the concept that living systems possess some immaterial, indescribable "vital" essence—emerged because this crucial distinction between information processing and physical substrates was not clearly understood. Historically, vitalism emerged in some form in almost every culture and usually in association with prescriptions for life and medical treatment. In the West, the invention of the microscope in the sixteenth century started vitalism's fall from glory, and the flowering of chemistry in the nineteenth and early twentieth century pretty much showed it to be wrong. Despite demonstrations that chemical processes in cells were just like those in laboratories, living systems still seem so much—well, "livelier" than mere chemical systems. Living systems seem to have richer repertoires of behavior, so perhaps this contributed (and may contribute today) to the idea that something was different. Something was indeed different, but it's the information processing that's different, not the components. My sense is that early on, without Turing's guiding ideas about computation, it was difficult to articulate how a complex collection of physical interactions expressed by a cell could be considered to "live." There were, of course, belief-based motivations behind the idea, but these only explain the vigor behind the arguments for vitalism. This same distinction between parts and the information processing they support will be important throughout the book. But let's be clear, this idea about computation is no longer particularly radical—it now pervades biology: Structure and function are really information processing being implemented by the physical and chemical properties made available by biological molecules, cells, networks of cells, and so forth.

This is the wellspring of a coming scientific revolution, the synthesis of many levels of scientific inquiry catalyzed by computation and computational tools. Signs of the revolution abound: computational fluid dynamics, computational learning theory, computational biology, bioinformatics, computational quantum electrodynamics, and on it goes. Computation is an essential tool in every quantitative science. But this is just its use as a tool like a microscope or a thermometer or a hoe. In biology, computation is also an indispensable conceptual tool, one that separates clearly the information-processing parts of a problem from what we used to call the physical or mechanistic parts and interactions. The second major symptom of the coming revolution is the ongoing dissolution of traditional academic and departmental boundaries. Currently, this is occurring at the fertile intersections of fields, those places where disciplines overlap—in the best cases, where they collide. In the twentieth century, the wholesale importation of physics, chemistry, and engineering into biology transformed it, invented what is now molecular biology, and set off what can only be described as a biological revolution.

But in the twenty-first century, the wholesale importation will come from the computational sciences and those areas of mathematics, physics, and even economics that impinge on the nature of computational descriptions. The signs are all around. Let's return to the example of DNA. It is now understood at many levels; its physical and chemical properties can be quantitatively described, but beyond that, animals can be cloned, genes can be constructed and used to build novel organisms, and sequencing is carried out by sophisticated robots. The list of amazing biological acrobatics will continue. My assessment here is mild compared to some—consider popular author Ray Kurzweil's prediction that machines will become sentient in the next thirty years or so. I hope my deskside computer doesn't get wind of this; I regularly spill coffee on its case with no apologies or remorse. Whether or not Kurzweil's vision is correct in detail, his sentiment is right on the mark, because the underlying revolution is on the march. Computational scientists in the coming century will address and begin to connect phenomena from the quantum mechanical to the mental—I suspect that it's not too long before it will be impossible to identify any parts of science not completely dependent on computational descriptions. Turing's original insight is as singular as Darwin's idea about natural selection, and like all great ideas, its simplicity hides its depth.

The computational theory of mind explains why thoughts require a properly operating brain to exist, yet it also shows why they (our thoughts) are not equivalent to the brain or its parts. It's the information processing that the brain carries out that is equivalent to our thoughts, not the parts themselves. But what about the meaning? The stuff the philosophers claim correctly is missing from the computational theory of mind? It is still missing in my description so far and will require another idea.

IT HAS BEEN SAID about Darwin's theory of evolution that it's the ultimate tautology—the survivors survive. This faux complaint presents a powerful feature of evolution; whatever works to keep you alive and get your genes into the next generation is just fine, no matter how weird the reasons or the results. But there is another feature of survival that is often not emphasized so much—survival is hard, desperately hard. Darwin understood this clearly and emphasized it in his title to Chapter 3 in *The Origin of Species*—"The Struggle for Existence." Mere persistence from one moment to the next is a struggle. So the augmented tautology becomes "The survivors survive but their life is desperate." Desperation is easy to envision for the life of bacteria; scurrying about trying to collect energy and reproduce, eliminated by the millions by random events like a puddle drying out or some animal urinating on them. Desperation is perhaps more difficult to envision for modern humans embedded in a culture that overproduces food and where at least one large nation's major health problem is obesity. However, all early humans can be seen as quite desperate, living a hard life with the threat of starvation as a constant motivator. Hunting and gathering is simply not very efficient, and until agriculture was discovered, early humans were always just one major mistake away from starving to death. This point is hard to overemphasize. Life is unforgiving, and so life's mechanisms had a constant pressure to be efficient—to capture, store, and process energy efficiently. And when we look at the components of life, cells, they are literal wonders of efficient energy-handling.

Out of the pressure of desperation comes efficiency. We all know that we become much more efficient and creative when we are desperate—when circumstances dictate that we absolutely must find some solution to a problem even though time and money have almost run out. Desperation is indeed the mother of invention. Plato called it necessity, but he really meant desperation. Life itself responds the same way: The tougher the times, the more crafty and efficient the so-lution; such is the power of evolution. What exactly is efficiency here?

Efficiency = the best long-term returns from the least immediate investment

Just like the "buy low, sell high" tip. But to be efficient, a system must value its available options, and valuation means estimating two things: the costs and the long-term returns associated with each option. Consider the following example.

You go into a supermarket, intent on buying milk and bread. You have a fixed amount of money, but lots of options would solve your problem—milk brand A and bread brand B; or perhaps milk brand X and bread brand Y. The average supermarket would present hundreds of possibilities. But this store is bit defective; all the price tags are missing. Now you are stuck. Without price tags, how can you tell the most efficient way to spend your money on these two items? You can't. Although you know the long-term benefits of eating each brand of milk and bread (let's assume you do), you cannot make an efficient selection without knowing the cost of each alternative pair of brands.

You go to another supermarket, again intent on purchasing just milk and bread. Thankfully, this store has price tags. However, now you have no knowledge about the long-term benefits of consuming one brand of milk or bread over another. Although you know all the costs perfectly, you still can't decide what to do. Perhaps the cheapest brands yield the same long-term benefits as the higher-priced brands, but you just don't know.

There is now quite a chasm to cross. How do these very mechan-ical- and economic-sounding ideas of computation, efficiency, and valuation relate to a human's ability to set and pursue goals, to trust someone, to make decisions that deny instincts, and so forth? The surprising answer is that efficient computations care—or more precisely, they have a way to care. I know that this sounds strange. And what exactly does an efficient computation care about? Goals. A machine with goals has implicit within it a natural measure of what its component processes mean; that is, the degree to which it should care about them. Let's unravel this idea.

There is always more than one way to solve a problem—any problem. If the problem is moving a brick pile from one side of a road to the other, then one solution is to move it all at once in a dump truck. Another solution would be to use a handheld bucket and carry the entire pile, bucket by bucket, across the road. Which is the better solution? The answer depends on two factors, cost and return, investment and payoff. The choice that returns the most for the least should win.

Computational problems faced by biological organisms present a similar trade-off, because to every problem there is always more than one solution. But all computations are not created equal. Some cost more to run, and some provide better long-term payoffs to the organism. For biological computations, efficient solutions have won the competition. How do we know? Because your brain is merely warm—you can safely touch your head—while the processor in your personal computer is so wastefully hot that it heats your office and you can't touch it with a bare finger. Why is the brain so efficient? The why is obvious: Life is hard and competition fierce, so biological computers could never afford to be grossly inefficient like our personal computers. But our question is, how do biological computers achieve such efficiency?

Remarkably, the answer to this question will connect low-level efficiencies in the operation of your brain to psychological mechanisms that allow us to trust, to love, and to deny our instincts. And here's the guiding idea:

> Nature has equipped biological computations with a measure
> of their value.

In biological computers, computations are not lifeless streams of symbols, totally devoid of meaning. Instead, biological computations carry something extra—an extra measure of their overall worth. Instead of just the computation, there's the computation plus "something else," and that "something else" is a measure of the value of that computation to the overall success of the organism, its overall fitness. It is as though every computation is paired with an assessment of the likely long-term return to be had should the organism choose to run the computation. It's like equipping every small piece of your car with something we might call value tags, which rank the overall value (to the car) of each part. Gaskets, screws, valves, suspension, flecks of paint, onboard computers, their algorithms, and so on are all equipped with a number ranking their relative value for the overall operation of the car. A reliability engineer would love to have such tags when assessing the performance of some engine or manufacturing system. This is one essential way that biological computation differs fundamentally from simple symbol manipulation.

So valuations put meaning back into neural computations. I will explore the different ways that the brain assigns such values, passes them from one symbol to another, and updates them based on experience and even simulated experience. The discussions will stay mainly at altitude, focusing on verbal descriptions of the value computations and how they are used to guide behavior. However, each area touched upon is a major area of inquiry in neuroscience, so don't be fooled by the lack of discussion of our knowledge of the underlying physical interactions, brain circuits, and molecular networks involved. This information exists and is being uncovered, categorized, and understood at deeper and more detailed levels. Our journey is like that of a water bug—skitter along at the top of this growing ocean of physical knowledge to get a glimpse of the computations.

AFTER TURING PUBLISHED HIS famous 1936 paper on computation, he moved to America to study with mathematical logician Alonzo Church at Princeton University. But world events rapidly intervened and in 1939 Turing found himself at Bletchley Park, a mansion just outside London that became Britain's code-breaking headquarters throughout the war. He was soon at the center of British intelligence's fight with the two brains, one cold and one warm—but both wreaking havoc in the North Atlantic—the Enigma machine and Admiral Karl Dönitz. The story of the effort to break the Enigma machine codes is fascinating and worth reading, but here I want to point out a critical feature of that effort—urgency. At Bletchley Park, Turing built on early work by Polish mathematicians and ideas from colleagues to produce a series of electromechanical devices called bombes for computing (breaking) the Enigma codes. The most difficult codes were those used by the German navy, including the U-boats. But if Turing was to save lives, speed and accuracy would need to be his twin goals. Break those codes—now. Energy usage and its distribution throughout the computing machine were not important. Enormous resources were poured into this effort because without it Britons feared the loss of truly valuable resources—their freedom and their lives. One re-source, efficient use of energy, was traded for another.

It is widely thought that the effort of the group at Bletchley Park saved many lives during the war. But I think that it also attached to modern computing an odd legacy. The model for the code-breaker was speed and accuracy at all costs, which means loads of wasted heat. The machines did not have to decide how much energy each computation should get; instead, they simply oversupplied energy to all the computations and wasted most of it. And although we under-stand the urgent needs of the time, this style allowed them all to over-look a critical fact—the amount of energy a computation "should" get is a measure of its value to the overall goal. Goals and energy alloca-tion under those goals; these two features are generally missing from the modern model of computing today. Just as for the code-breakers, speed and accuracy are the primary constraints on modern comput-ers. Run chips faster and faster and make computations more accu-rate so that simulations of planes, fluid flow, atomic vibrations, elevator schedules, bridges, and flight paths can be more realistic and more useful. And while it is true that accurate computations are usu-ally needed for such jobs, for truly adaptive computations, those that use energy wisely and can adapt to that finger slip on the mouse or that unexpected button press, another radically different route is re-quired. And using energy wisely is the basis of why biological compu-tations know how to care.

The best example of a computational device that handles energy wisely but still carries out unimaginable feats of intelligence is not far away. Just query the human brain; it's almost perfect: It's slow, noisy, and imprecise.

THE BRAIN IS (ALMOST) PERFECT

It's Slow, Noisy, and Imprecise

> Recharge or die; this is
> the first law of efficient computation.
> ANONYMOUS

THE BRAIN IS A set of living computational devices that generate our mind, and it's almost miraculously efficient. How do we know? We know because the brain possesses all the characteristics of a highly efficient computational machine—it's slow, noisy, and imprecise. And one more symptom: It has goals. Huh? Does that sound like any computer in your office? Our brains are vastly better than anything Silicon Valley has produced. Slow, imprecise, and noisy; it's hard to see how the human brain could get any better. Go with me on this.

Let's start with the Energizer Bunny; they say she just keeps on going and going. This pink, shade-wearing bunny is the cultural icon of constant motion, long-lasting effort, and better batteries. Her ability to influence decisions about what kind of battery to buy derives from what she represents to brains familiar with her. For now, let's focus on a secret about the Energizer Bunny that we all know but don't talk about. She won't keep going. She'll eventually stop moving. There will be no Energizerness left; only an immobile, pink bunny. Even if she runs longer and stronger than others of her kind, if her batteries are not recharged, then we all know that the motion will stop. She will eventually succumb to the real principle of moment-to-moment survival, recharge or die, the first law of efficient computation.

Survival is indeed hard. Why? Because the economic realities of life never ease up. Living is not really an uphill climb with hilltop awaiting; instead, it's a treadmill—lots of effort just to stay in place and not be thrown into oblivion (or the gym floor). Mistakes on life's treadmill are not easily forgiven, and so mobile creatures must invest their choices with care. They run on batteries and they must continually acquire nutrients and expel waste in order to recharge their batteries and live another day. If you don't eat, you don't survive. If you don't survive, you don't reproduce. If you don't reproduce, well, you know the rest. Prey are batteries to be drained by predators. There is no free lunch; everything is quid pro quo, and so the name of the game is recharge. This doctrine mandates that evolution discover efficient computational systems that know how to capture, process, store, and reuse energy efficiently. Organisms that manage energy efficiently will do better than those that don't. And organisms with efficient computations will enjoy the same benefits. The need for the efficient handling of energy for fuel and for information processing has constrained life ever since there was life.

These observations about moment-to-moment survival generate a list of counterintuitive symptoms of the features expected in efficient computational devices. These symptoms will expose a myth that has built up in the world of neuroscience. I'm not sure where it originated, but like the whisper game in kindergarten, the initial message, whatever it was, has morphed into a kind of mantra about the seeming sloppiness of biological computing. There is no evidence that biological computation is sloppy. It is typical to read or hear descriptions to the effect that, despite the "slowness" of neurons, the "clunkiness" of their parts and interactions, and the "noisiness" of the communication that goes on, the brain somehow (usually through the miracle of parallelism) produces marvelously subtle perceptions and behaviors that no man-made machine can now rival. In this tale, slowness, imprecision, and noisiness are liabilities. I admit up front that these features are indeed liabilities to fast and accurate computation in a machine with excess internal communication channels where cost is not an issue. While a machine willing to waste energy ("get hot") on computation speed, accuracy, and internal communication may well best the human brain on specific tasks, this misses my point. An F-15 fighter jet flies faster than any bird, but a bird can maneuver through trees, land on branches, and make copies of itself—all without jet fuel. If we are serious about building and understanding computational systems that can deal with the real world, then we must remember that creatures run on batteries that they themselves must recharge. Once costs are added back into ideas about computation—once efficiency is considered—these biological liabilities become exactly the features we should expect to have evolved.

The style of computation used by the brain is very different from that of the computers inhabiting our cars, offices, and server rooms. The energy efficiency of operation for the entire human body is staggering. The average hundred-watt lightbulb costs about a penny an hour to run, at average market rates for electricity in the United States in 2005: around ten cents per kilowatt-hour. Go ahead and check your monthly bill. A human being sitting comfortably in a chair consumes energy at a rate of about a hundred watts, roughly equivalent to the average lightbulb! And this consumption is running literally everything—digestion, blood pumping, breathing, mental function, and a myriad of other processes. The brain consumes about a fifth of this rate; therefore, while sitting, the brain costs about a penny every five hours to operate, less than a nickel a day—now, that's an efficient machine!

This argument does not count the incalculable expense (over generations) of actually evolving humans and building a particular human complete with experience, but it does show that once built and running, humans are freakishly efficient thinking machines. There is no way around this conclusion. No matter how one allocates this small energy consumption to different brain functions, the conclusion is unavoidable: The brain is an extremely efficient computational device. Not so for the computer by my desk. In fact, it makes my office quite hot, and if I remove the cover and touch the central processor, I will be badly burned (don't try this). The processor is producing a lot of wasted heat in the course of turning power consumption into useful computations, including those that underlie the programs we all use—e-mail, word processing programs, spreadsheets, Internet browser, and so on.

Efficiency is a concept common to discussions about evolutionary adaptations. All things being equal, an organism that efficiently collects and uses energy is much more likely to survive and reproduce than one that does so less efficiently. This idea is a special case of evolutionary fitness, the capacity of an organism to transmit its genes into the next generation. I could trace the evolutionary adaptations that have arisen to capture and utilize energy in biological systems and the way that these adaptations contribute to overall fitness. This is an extremely interesting topic, because it's about some remarkable nanoengineering feats. Humans still have an enormous amount to learn about nanoengineering from biology, and most institutions devoted to these issues are stealing ideas from biology and in some cases constructing systems that use biological components in conjunction with human-engineered components. This direction would take us through arguments about various adaptations and into areas of evolutionary biology and evolutionary psychology. I have chosen to confine my discussion to how things are done now, how decisions are made moment to moment. I cannot avoid talking about evolutionary adaptations, but my focus will be on our best understanding of how things are accomplished in a working human brain.

WHAT IDEAS TRANSFORM imprecision and slowness and noisy communications from liabilities into features of neural function? I will answer this question by introducing principles of efficient computation and use these to organize our understanding of mental function and even mental disease at three different levels: (1) neural hardware (structure and function), (2) mental models run by the brain, and (3) social instincts for trade and exchange. These levels meet on the common ground of computation, and I investigate them using the disciplines of neuroscience, psychology, and behavioral economics. The principles of efficient computation will reveal how our nervous system can set and pursue goals, can assign value to our thoughts, and can build efficient models of other people. But first we must deal with the ugly reality of costs.

Suppose that you are an engineer and some agency asks you to build a chemical plant to produce a new fuel-cell technology for cars of the future. When confronted with such a request, an engineer asks two questions: What do you want the plant to do? ("What do you value?") And what's the budget? ("What does it cost?"). In a hypothetical world with unlimited time and energy, the engineer might just ask what kind of optimal product the agency was interested in producing. It could take many generations to discover the correct answer to the design specifications and several more generations to build it, but since time doesn't matter, this would not be a worry. I'm not sure where such luxury might actually exist, but I know that it's not on this planet. And even with such a luxury, the answer might be disappointing. This feature of languid decision-making was recognized by the late science-fiction writer Douglas Adams in *The Hitchhiker's Guide to the Galaxy*, where a hyperintelligent race builds the second-greatest computer ever, called Deep Thought. Although the computer was only the second-greatest ever, the designers did not lack for ambition and they asked it for ". . . the ultimate answer to life, the universe, and everything." Tall order. Deep Thought takes its time, lots of time, but, after seven and a half million years and many generations, finally emits the answer: "42." Needless to say, the hyperintelligent designers of Deep Thought were sorely disappointed. I personally found the answer disappointing—forty-two is not even a prime number. The hilarity of Adams's imagined world shadows the fact that even unlimited time and energy is not the answer to optimal design. Without costs, the question of optimal design attacks itself and becomes "Optimal for what?" In the real world, these luxuries of time and space and energy do not present themselves and everything has a cost. Consequently, one cannot profitably ask how the brain works and how this connects to psychology without including a consideration of costs. We might not know how to measure costs in every circumstance, but the extra step of contemplating costs produces principles that unite neuroscience, psychology, and economics under a single efficient computation imperative: Compute the answer as efficiently as possible given the imposed costs.

A COMPUTATION IS THE manipulation of any set of symbols according to a set of rules. Basically, anything that meets this description could in principle be simulated on a computer, maybe not effectively, but simulated nevertheless. As a child in grammar school, computations were "those things we did in math class," which included problems like multiplication, division, and algebra. It was all pretty straightforward then. For example, we were asked to replace "2 + 1" with "3" or "15 × 3" with "45." These simple operations make the point—transform one set of symbols into another according to prearranged rules of manipulation. There is an unstated cost here— we were all given the same amount of time to come up with the answers. But the notion of computation does not end with simple calculations from school. Rather, it can extend into every domain of our deliberative mental lives. A human that reads a musical score and translates it into a series of finger movements is carrying out a computation, albeit a complex and subtle one. One set of symbols (notes on the page) is being changed into another set of symbols (finger movements). The finger movements (keys struck) are changed into pressure variations in the air (sound waves), picked up by the ear, and changed into patterns of electrochemical activity in the brain. Each of these steps is a complex set of computations. Turning a series of experiences and the feelings they engender into a story or poem can also be understood as a computation. I want to be careful here to differentiate my claims. Although this is a remarkable insight for the sciences of mind, it does not explain the total experience of poetry. It explains how something as subtle and complex as poetry could run on a human brain. Poetry and music are still capable of engendering experiences in us that we cannot capture in computational models. I'm speaking here about the complex, but easier, parts of the problem. Even our understanding of physics can be viewed as computations—compressed models of experience, encoded into mathematical descriptions, and capable of reproducing (predicting) the outcome of experiments.

So poetry, physics, music, and mental arithmetic can be captured by computational descriptions. Unless something magic is happening in our brains, this is the best prevailing and testable hypothesis about how a biological organ produces interesting cognition that we care about. With the idea of computation restated, let's now ask two questions: How can computational devices save energy? And are there clear symptoms that show that a computational system is designed for efficiency?

FOR OUR PURPOSES, I first need to distinguish between free energy and entropy (wasted heat). Free energy is the good stuff. It's the energy in a system that's available to do work ("free to do work") like the energy that turns the sail on a modern windmill or the energy that causes the piston to rise in a car engine just after the fuel-and-air mixture explodes. However, all the energy doesn't turn the sail or push the piston. Some of it is wasted and can't be recovered to be put to other useful work. When the sail turns, friction causes some of the energy to be lost as unrecoverable heat. When the fuel explodes, the piston moves, but it, too, produces unrecoverable heat through friction and even through the production of sound waves in the air or vibrations in the engine block. The idea of recoverable and unrecoverable energy can be made intuitive by counting the number of ways that energy can be transferred. Let's carry out a simple thought experiment.

Push an imaginary billiard ball gently across an air hockey table so that we can ignore the friction with the table and the surrounding air. On the way over to the other side, the ball touches a paddle connected to the table by a spring and starts the paddle rocking back and forth. When the ball arrives at the other side, we notice that it is moving slightly slower. Some of its energy has been transferred to the wobbling paddle. In principle, we could recover this energy by connecting the wobbling paddle to some device that it could pull or push; maybe connect the paddle to a little generator that makes electricity and charges an attached battery. Here's the sequence. The paddle steals a little energy from the moving billiard ball, starts moving back and forth, and we could recover that energy. Now imagine that we have two paddles connected by springs and they both are touched by the ball as it moves across the table. Again, the ball slows and its energy of motion is transferred to movement in the two paddles. Since there are only two paddles, we can easily extend this thought experiment to include the recovery of their energy. Now let's change this scenario into a real pool table covered with felt. Each fiber in the felt acts just like the paddle. When we push the cue ball across the table, it touches millions of tiny fibers in the felt, they vibrate a little, and we again notice that the ball is moving slower at the other side of the table. The energy from the moving ball has been bled away by the fibers. But it's worse than this. The ball also bangs into trillions upon trillions of oxygen and nitrogen molecules in the air and they, too, carry away some of the energy. The problem here is a numbers game; it's statistical. There are too many ways in which energy can bleed away into unrecoverable forms. Each distinct way in which energy is lost would have to be coupled to a means to recover it, and the numbers quickly become overwhelmingly large. We could never hope to capture this energy again, so it's lost for good and never comes back to help us do work—it's wasted heat. Car engines produce a lot of wasted heat. They stay hot, radiating heat into the night air after being turned off. If they were perfectly or near perfectly efficient, then they would be cold to the touch very quickly after the car stopped. The brain does not waste heat—it could never have afforded to do so, and this is crucial to how we choose.

FOUR SAVINGS PRINCIPLES guide the way that we characterize efficient computational machines. Many of them might also readily apply to savings accounts. They are not altogether independent, since they're all just "conserve energy" strategies; however, they predict many features to be expected in any efficient computing machine, including one built out of biological parts. They also prescribe the need for an efficient device to have goals.

> **Principle 1.** Drain batteries slowly.
> **Principle 2.** Save space.
> **Principle 3.** Save bandwidth.
> **Principle 4.** Have goals.

What should we expect from this prescription? Let's examine the kinds of features that each savings principle predicts and compare these predicted features to the known features of the mammalian brain.

Principle 1. Drain Batteries Slowly. Expected consequence: Compute as slowly and "softly" as possible.

For any amount of energy an organism can store (its battery life), all things being equal, it's better to consume as slowly as possible, since it drains the batteries at a slower rate—one never knows when that next recharge will come along. Two related features follow from the "drain batteries slowly" principle. Compute as slowly as possible: Faster rates of computation consume more energy per unit time. What exactly does it mean to compute as "softly" as possible? In a personal computer, charging and discharging capacitors (specific circuit elements) is one way that information is stored and retrieved. However, all charging-discharging transitions are not created equal; "spiky" transitions waste more energy than soft, gentle transitions. Let's use semaphore as an example. Semaphore is a visual method of communication where the sender uses a flag in each hand and spells out words using different arm positions for each letter. All things being equal, the higher the number of letters sent per unit time, the more energy is spent. However, for any fixed rate of information transmission by semaphore—let's say one letter every two seconds—some semaphore styles waste more energy than others. The sender will save energy if the transition from having "arms at side" to "arms in positions that represent the letter *a*" are not fast and jerky, but instead slow and gentle, yet still within the allotted one letter every two seconds. This means that the sender wants to spread out over the allotted time his transitions from one letter to the next. The slower-gentler approach, the soft semaphore, requires less energy per letter.

Our current styles of computing machines completely ignore both savings measures and instead run computers at great rates and use very "spiky" transitions. Consequently, they create a lot of wasted heat. Biological computers are wondrous handlers of free energy (energy that can be converted to useful work)—they can store it in chemical bonds, recover it, and pass it around from place to place with high efficiency. This is a major reason that our brains can run on twenty watts without the blazing temperatures produced by our desk-side machines. A biological organism can compute only as slowly and gently as is consistent with its own survival. The "drain slowly" principle shows that slowness in computation should be seen as an adaptation, not a liability to be surmounted.

To see if there is neural evidence for slow and soft computing, let's start with some basic observations about the nervous system. The most important single fact is that it is made of cells. Building the computational machinery of the brain out of cells is a crucial fact, because cells already know how to process free energy efficiently, to replace and repair their parts on demand, and like all other cells in the body, they contain a complete copy of all the genes required to build a human. Furthermore, the brain has sophisticated power and raw-material distribution systems in the form of a complex system of blood vessels that listen very carefully to the needs of the surrounding brain tissue. Whenever cells in the brain need more resources, they send signals to nearby blood vessels and this resource request is met by a local increase in blood flow, which delivers oxygen, glucose, and other essentials. Those are some fancy fundamental building blocks!

There aren't many computer chips whose parts can repair themselves when needed, ask and receive energy as needed, and contain a blueprint that could in principle rebuild the entire computer. There are two main cell types in the brain, neurons and glial cells. Neurons possess unique structural specializations called axons, dendrites, and synapses. Axons and dendrites are the wires of the brain, and since the cell already knows how to repair itself, so do the wires and the connections (synapses) they make. These wires are really thin; a cubic millimeter of brain tissue contains about a mile of axons and a couple of miles of dendrites. So the brain has miniaturization and self-repair to add to its bag of tricks.

The primary mode of long-distance communication from one neuron to the next is through messages encoded in patterns of electrical impulses that move along axons and dendrites. These impulses are called action potentials: deflections in the voltage across the neuron's membrane that travel like a "snap in a garden hose" along the axon to the synapses at its tips. These neural "snaps" provide strong evidence for slow and soft computing: They move slowly, they have a long duration, and they are produced at low rates.

In the cerebral cortex, speeds of transmission range from one to thirty meters per second along axons and around one third of a meter per second along dendrites. These impulses are really taking their merry time to get to their destinations. Compare these speeds to the near speed-of-light transmission times along the "wires" in a personal computer. Light travels at about three hundred million meters per second, so there is about a thirty-million-to-one ratio between brain communication speeds and personal-computer internal communication speeds. In the nervous system, to make the neural impulses travel faster, significantly more energy would have to be expended by the axon (the wire).

The neural impulses are not just slow in transit, they are also luxuriously wide. Each neural "snap" (impulse) is about one to two thousandths of a second in duration—this is the total time for an impulse to rise from its resting value to its maximum and back to its resting value. This may seem fast in human terms, but it's miserably slow in computing terms. The electrical impulses that encode ones and zeroes in your personal computer are about a million times faster. Although this speed sounds impressive, and it is in one sense, it's also terribly wasteful. The faster the rise and fall of an electrical impulse, that is, the "spikier" the impulse, the more wasted heat is generated. A truly efficient machine would make the rise and fall of the information-bearing electrical impulse as gentle as possible. It's technically difficult to say whether neural impulses are "as gentle as possible," but they are many orders of magnitude more gentle, and hence more cost-saving, than the impulses that carry information in modern computers.

Lastly, neurons produce action potentials at a leisurely rate. Of course, we might have guessed this by observing the sheer fatness of the impulses. Since one to two thousandths of a second are required to even "define" an action potential, we see that a neuron could not possibly produce more than around a thousand action potentials per second. Indeed, this is the case, and most neurons fire away at around twenty to one hundred times per second and can accelerate up to several hundred pulses per second, but only for very brief periods. Some exceptions exist in systems that control fast eye movements, but this is a good ballpark figure.

We see from the discussion above that neurons produce slow-moving, soft impulses at leisurely rates—all symptoms of energy-saving maneuvers. Does this mean that the neurons in our brain compute slowly and softly? Yes. This conclusion is further supported by the fact that the main mode of communication, action-potential movement along the wires (axons and dendrites), is also thought to be one of the primary modes of computation in the brain. So Principle 1 gains real support from some of the most basic functional features of neurons. Before moving to Principle 2, I should add one thought about this description of neural computation: It's extraordinarily simplified.

Although I freely use the terms *wires* and *connections*, these metaphors are only meant to indicate the general roles played by axons, dendrites, and synapses. They are all complex, highly adaptive structures. Synapses and dendrites can change their function dramatically, depending on the impulse-encoded messages they receive. They can also change their function depending on chemically encoded messages from surrounding regions of neural tissue. These changes are thought to be one way that these structures flexibly direct the storage and recall of information in the brain. In addition, synapses, dendrites, and axons all contain highly interconnected biochemical networks with a vast computational capacity that is very poorly understood. So there is a lot more computational machinery than we have skimmed here, it's just not very well understood yet. We will continue to mention only those features of neurons that serve our discussion while remembering that the biology of neurons and their connections is an immensely large and complex field of study.

Principle 2. Save Space. Expected consequence: Be as imprecise as possible and compress everything.

More space consumes more resources, even if it is just storage space. A larger warehouse costs more to clean, repair, and organize. We would expect an efficient computational device to minimize the number and size of its parts in a fashion consistent with its goals. The first computational consequence of this principle is to compute as *imprecisely as possible*. More digits per number cost more to process and store than fewer digits per number, so if the machine is representing numbers in any way, it will seek ways to reduce the required precision. *Minimize components* means to shrink the physical size of components and reduce the number of components for any given task. Use as few distinct parts as possible. This requirement holds as much for a physical machine as it does for an algorithm that specifies a number of computational steps to accomplish some goal. All things being equal, algorithms with fewer distinct steps are more efficient than those with more distinct steps. They are smaller and they are less likely to produce errors. It's much more efficient to cook microwave-style pot roast and vegetables—"puncture top of package, place in microwave, and cook on high for three minutes"—and this procedure fails only rarely. The analogous "pot roast and vegetables from scratch" algorithm is long and littered with all sorts of difficult steps that invite mistakes—you get the point. And you probably have guessed that I'm not much of a cook.

The "save space" principle carries a powerful implication related to the style of computing of an efficient computational device. *Saving space* means compression. Compress the size of the parts. Compress the number of parts. In particular, compress the number and size of communication channels whether or not wires are used to implement these channels. Wires are a particularly costly form of communication channel and their total length must be minimized in an efficient device. Very interesting consequences flow out of these needs for compression.

To my knowledge reverse origami is not yet an official Japanese art form, but it has always been lurking around the edges of origami. Origami is the art of folding paper into beautiful three-dimensional objects like boats, birds, and other wonderful shapes. Birds are my favorite. I am most amazed when a lush three-dimensional shape springs forth from just a few folds in the paper. Reverse origami is the inverse of this process and is part of the thinking that goes on in the origami artist's brain. The idea is to observe a beautiful three-dimensional object and, through some series of mental steps, arrive at a way to represent this as folds in a single flat piece of paper. This reverse perspective shows that, in addition to artistry, origami contains the concept of a compression algorithm. Every three-dimensional object can be represented by lines drawn on a flat page (the folds) and a few extra instructions for the order of folding. If we wanted to store a menagerie of three-dimensional origami animals, we could pack them into a container or we could simply store a dramatically smaller box of papers with lines on them. The box of papers would take up much less space than the menagerie.

One could compress this paper representation even further. Lines drawn on a piece of paper can be represented as set of "drawing instructions" in an appropriate programming language. So the three-dimensional animals could be compressed to a two-dimensional page or further compressed to a short sequence of one-dimensional computer code, which takes up very little space. In this way, a collection of ten thousand origami animals could be stored on a USB flash drive (key-chain drive) and carried around in someone's pocket. This shows how something complex like a three-dimensional animal sculpture might be represented as an extremely compressed collection of one-dimensional instructions.

Indeed, origami is compressed and transmitted all over the world in exactly this fashion. Google "origami" and you will find numerous Web sites for it; click on a site and you can download page drawings with folding lines as well as pages of folding instructions. There is a crucial missing element left out of this discussion—the folder. That's right, the "reader" of the compressed instructions, the person who must be able to follow the instructions and execute the folding correctly. Origami instructions make deep assumptions about the capacities of the entity that will execute them, that is, the person (or device) that will uncompress and follow them. In this sense, origami is a message sent to someone, and each end of this communication must have a model of the other. The sender must draw the fold lines in a way that is comprehensible to the intended receiver and the receiver must know what the sender intended. Without such models, the origami instructions are meaningless. In this sense, decompression is the act of modeling the source of the compressed instructions. A recipient who does not have a correct model would be helpless. If you give your dog the origami instructions printed on clean paper, three-dimensional paper sculptures will not result. I tried this experiment, and my dog put several scratch marks on the paper and licked it; no sculpture emerged.

Now that we see the relationship between compression and modeling, let's put a few more cards on the table—all compression is not created equal. Our origami example helps. We have two choices for compression schemes, lines on a sheet of paper (folding instructions) or some computer code on a USB flash drive. Each scheme represents a different decompression cost to me (I'm the recipient here). The USB flash drive version requires that I have a computer, the right program, and a printer to print out the folding instructions on a piece of paper. After that I can fold the shape. Although this is an extremely compact way to represent the paper sculpture, it may be too costly for me to uncompress. The moral of this story: The smallest representation is not always the best in the presence of limited resources, that is, all real-world situations.

This problem suggests the need for "just in time" compression, where data is compressed to a level where the decompression can occur "just in time": not too early and not too late and depending on the cost that the decompressor is willing to pay. If some object is decompressed too early, the information must be held in some kind of temporary storage (buffer), and this has a cost. If it is decompressed too late, then what's the point? Of course, everything depends on the nature of what's being compressed, but there is a more profound point here. For any specific data, the depth of compression will depend on a trade-off between the cost of decompression time and the cost of buffering. There will be some cases where data (or programs) could be further compressed by a particular computational device but aren't because the costs of decompression are not worth the extra space savings.

THE MANDATE TO SAVE space prescribes two structural features for machines that use wires to communicate: (1) Build as few wires as possible and (2) build more short wires than long wires (conserve wire). Indeed, the neural symptoms of the "save space" command can be seen most clearly in the wiring strategies adopted by the brain, and in particular the cerebral cortex. There has been substantial theoretical and experimental work on this topic directed at the cerebral cortex of mammals or the entire nervous system of much simpler creatures. In the mammalian cortex, "minimal wiring" ideas have been used to analyze connection strategies observed there. These include connection strategies among different brain regions dedicated to specific processing functions (e.g., vision), connection strategies within single brain regions (e.g., primary visual cortex), connection strategies within modules that make up single brain regions, and connection strategies employed by single neurons. A consistent "minimal wiring" picture emerges, although every rule is better seen as a statistical trend and not a "law of connection-making."

For example, single neurons use most of their axonal length making connections nearby; they make fewer intermediate-range connections, and even fewer longer-range connections. In the particular case of long-range connections, the cerebral cortex is sparsely connected; one site connects to only a small fraction of all the other sites. It is thought that the cortex can get away with this sparse long-range connection strategy by using lots of shorter local connections in and around the targets of these longer connections. This is a common theme. A neuron will often make a long-range connection that expresses a lot of little branches only near the target site, rather than sending a bunch of separate branches over the long distance. This strategy saves on wire length, but it also makes copies of the messages using the local branches at the target. Axons are active devices all along their length, and an axon branch is functionally equivalent to a "copy" command—messages encoded as neural impulse patterns are cloned at each branch point of the axon and move out the daughter branches. The connectivity of the human cerebral cortex is immensely complicated, and my simple summaries do not do the field justice. It is a scientific area in which structure and function are almost artistically blended. Despite this complexity, there are indeed statistical trends that lend substantial weight to the idea of minimal wiring, which is itself a symptom of the need to save space—a necessary strategy for a machine that must carry its own power source and collect its own fuel.

Principle 3. Save Bandwidth. Expected consequence: Stay off the lines, don't repeat yourself, and be as noisy as possible.

Bandwidth is an engineering term that refers to the "size of the pipe" available for communication. The pipe metaphor helps here. Suppose we were putting messages into plastic balls and sending them down a water-filled pipe. The messages must share the same pipe; they occupy a fixed amount of space, and they take a fixed amount of time to process at the sending and receiving ends. There is a limited resource here: the maximum rate at which plastic balls with messages can be put in the pipe—the maximum rate of information transmission. This picture embodies the idea that all realistic communication channels have limited bandwidth. In the digital world, bandwidth refers to the same idea: the maximum rate at which bits of information can be sent. In my locale, the idea is familiar to everyone who has tried to browse the Web from home on Sunday nights. The response time is terrible; everything is sluggish. Loads of people are at home browsing the Web simultaneously. Too many message-bearing packets are being jammed into the same pipe and the system starts to slow down. It slows because there are too many messages to process and too few communication channels to share. The take-home point—the available bandwidth is a limited resource that must be used sparingly.

Limited bandwidth requires concrete strategies for sparing its use. We should expect to find these strategies in place in all efficient computational devices. The lowest-energy solution would be to use no bandwidth at all; never talk to the other parts of the device. This is practically impossible, but it generates the "stay off the lines" rule. What about those times when the machine must use its communication channels, those times when it's impossible to stay off the lines? If you must communicate, say only what needs to be said and nothing more. Try desperately hard not to repeat yourself. Repeating a message is one way to make sure that the receiver gets the entire message faithfully, but it's inefficient. Redundancy may protect against errors in transmission, but this style of error prevention has a cost. We have all experienced this issue with cell phones. We often must repeat a message because parts of it have ended up garbled, either because the connection is bad or because both speakers started speaking at the same moment and the two conversations collided. However, repetition consumes bandwidth on the communication channel and our precious cell-phone battery charge runs down just a little bit more than it would have if no repetition had occurred. Bad cell-phone connections cost us money. To be efficient we should keep such repetition to a minimum. How can an efficient device concurrently achieve these two goals of "stay off the lines" and "minimize repetition"? Modeling. The parts of the device that talk to one another should contain models of one another. This way, they know how to make intelligent assumptions about what their listeners are doing and expecting from them. A modeling strategy allows the machine to minimize communications between the parts and goes a long way to meeting the "stay off the lines" and "don't repeat yourself" commands.

What about when the communication is actually happening? Is

there a more efficient communication strategy during those moments when bandwidth must be consumed? Indeed there is, and it
sounds completely counterintuitive: When communicating, choose a
scheme for encoding messages that is nearly random! Technically,
the most efficient codes will look almost like noise on the communication line. Sounds a little crazy, but it's true. It's really not important
why this is true, but the take-home point is crucial for our efficient
computation symptoms list. If we were to eavesdrop on an efficient
communication channel, record the impulses carrying the messages,
and study them later, their pattern should look "nearly random"
through time.

A brief summary derived from our principles might read: "Go as
slow as possible, be imprecise, and stay off the lines. If you must talk,
don't repeat yourself, and in fact it's a good idea to be almost random." This reads like a formula for staying off the radar screen during
my childhood. "Don't make unnecessary fast movements, since it
alerts parents, be as vague as possible about what you are up to, stay
off the phone, but if you have to talk don't repeat yourself and speak
something like pig Latin." For me, this was a well-worn strategy for
staying out of trouble as a kid. These same general precepts also apply
to moving about in an area where there are predators that can kill and
eat you. The predator-prey version would read: "Go as slowly as possible for the moment, be efficient when you decide to do something
like run or eat, and don't announce yourself by making unnecessary
noises—lots of others are listening in on the line."

What would we expect out of neurons if they were part of a highly
efficient computational device trying to save bandwidth? We would
again expect low neural impulse rates in response to the "stay off the

lines" principle. Remember that low impulse rates also satisfied the "compute slowly" edict. And indeed neurons in the cerebral cortex fire at low rates, about twenty per second to about sixty per second. Sometimes they send out bursts, but these bursts do not last long (around ten milliseconds) and rarely exceed rates of four hundred to six hundred impulses per second. This is quite slow compared to impulse rates in a personal computer, which are typically in the range of billions of impulses per second.

The second neural symptom would be an apparent noisiness in the times of the impulses emitted by neurons. Exactly what do we mean by noisiness? Noisiness is easy to characterize. Suppose that I want to send a message using a communication channel and digital impulses that are either on or off (ones and zeroes). Also, let's suppose that I have chosen to limit the average rate of impulses that I will send to forty impulses per second. I might have chosen this slow rate to "stay off the lines" as much as possible. Now let's look more closely at a particular second of communication. During this time, I want to send forty impulses, and so I'm left with the forty decisions to make about the exact time of each impulse. Where should each of the forty impulses be arranged in our one-second window of time? A purely random selection of these forty times is possible using something called a Poisson interval distribution, and this prescription for choosing the forty times is theoretically the most efficient use of the bandwidth. This means that we should expect the collection of impulse times produced by a real neuron to follow this same Poisson interval distribution, if indeed the neurons are trying to save on bandwidth. Remarkably, impulses emitted by real neurons in the cerebral cortex and many other places in the nervous system look noisy in exactly this way. Instead of asking how neurons do amazing things *despite* the fact that they are noisy, we should marvel at how well they use the bandwidth made available to them. Again, what appears to be a liability is most likely an extremely optimized adaptation. This general perspective has been taken by biophysicist and neural theorist Bill Bialek of Princeton University. Bialek's approach has led to numerous insights into how the nervous system of any creature should and does encode and decode information. His work searches for optimization principles with which to understand neural function, and although he can't prove it in every case, everywhere he looks he sees biological systems working near the limits imposed by physics. The evidence that we have discussed so far tells me that he's looking in the right way and in the right places.

Neurons talk slowly, and when they do talk they talk using noisy codes. This is not a perspective normally taken when designing modern computers, since their usefulness to humans has always been measured by their ability to compute things quickly without regard to energy efficiency. Computing speed has been the reigning king. As we move into the twenty-first century, the need for pervasive, low-cost computing and the physical limitations for current computing styles will force us to reconsider the goals of our computational devices and the way they use energy. This raises the issue of the last of our four efficient-computation principles—goals. A computational device cannot be deemed efficient, either by some outside agency or by itself, without explicit goals. Three questions arise when we talk of the "goals of the machine": How does a machine have goals? What is the function of goals? What are the neural symptoms of goals?

Principle 4: Have Goals. Expected consequences: Physical signals representing goals, desires (error signals), and values.

None of these ideas about efficient computation carries any weight unless a computational system "wants" to do some things more than it "wants" to do other things, that is, it must have goals. Without goals, a computational system cannot be efficient for the simple reason that it has nothing against which to gauge its ongoing performance. The need for goals is a serious requirement for an efficient computational system, since goals provide crucial energy-saving feedback information. We have a strong intuitive sense of what goals are—they are the "ends," the "aims," the "state that is desired." Goals come in many shapes, sizes, and contexts: soccer ball in the back of the net, pay increase in the next fiscal year, new boyfriend or girlfriend, better job, better sense of self-worth, food for tomorrow's meals; the list goes on. How do goals serve efficient computation and how could a machine possess a goal? Let's turn to the squirrel that lives in my yard, since she makes the problem of abstract goals quite real.

I call it my yard because I pay the mortgage on the property; however, the squirrel doesn't know about mortgages and acts like she owns the place. She also acts like she has goals—many goals. My guess is that "Stay away from the large, growling creature" and "Hide nuts in places you can remember" are near the top of the goals list. Her actions reveal this. She hustles about hiding or digging up nuts and she keeps a wary eye on my dog. We could get much more specific about her behavior, but the point here is another question. How can her nervous system define these rather abstract goals for her? Surely there isn't some internal program that proclaims, "The growling, curly-haired, fast-moving creature should be avoided," and specifies all sorts of specific decisions to meet this abstract command? No, but something functionally equivalent happens through indirect feedback. *Indirect* is the key word here. Specifying complex goals directly is not an efficient way to build goals into a system. Our nervous system sidesteps the difficulty of directly defining complex goals. Instead, it comes equipped with a series of guidance systems that help us navigate complex terrains, decision-making problems, and social interactions. And they are efficient.

Suppose that I want to build a robot to achieve the goal "catch fish safely in the Alaskan wilderness." How could we possibly program this into the robot? We might rush into the problem and try to build complex decision-making algorithms around this global goal. Our robot-building project would require a lot of planning and knowledge about the difficulties the robot might encounter, and we would have to make some really good guesses about unexpected dangers and other surprises. We would encounter the same problems that faced the recent mobile probes sent to the Martian surface.

Instead of starting by defining the problem directly, we might instead sneak up on it indirectly. We could equip the robot with something akin to "guide" signals. These signals would function just like, well, fishing guides. If I take a fishing trip to Alaska, I need a guide because I don't know how to find the right rivers, the right fishing places, and I especially don't know how to avoid all the dangers Alaska presents, particularly the bears. But guides do. However, guides don't have a fixed rule book that they read and follow in order to "know" how to accomplish all these goals at once. Instead, they use their own rich mental models of "safe Alaskan fishing" that they have built up through experience and study, and combine them with new information as the particulars of a specific trip cause them to make small changes in plans. By combining stored models of what's valuable to do next with new information, they can help us achieve our goal of "fish safely in the Alaskan wilderness."

LET'S THINK ABOUT THE STYLE of direction that I would receive from a guide. Rather than giving me some gigantic download of life-long experience with the Alaskan wilderness, a guide will instead give what is best termed "local guidance." Where we will go next, what to look out for over the next hour, how and where to cast my fishing line, where not to go if I must excuse myself for a bathroom break. The guide will guide you, and the guidance will lead to a lot of little corrections in your current behavior and your behavior in the near future. In the best circumstances, all these little corrections will indirectly achieve the goal of "catch fish safely in the Alaskan wilderness." So, the instructions from the guide are like little navigation directives all collaborating together and indirectly to achieve a rather abstract goal. This is one way that the nervous system implements goals—it uses collections of corrective guidance signals (error signals) to navigate an individual's behavior. An explicit abstract account of the goal is not really necessary as long as the system produces and follows the guidance signals accurately. This means that many of the goals pursued by our nervous system will not be explicitly available as conscious experience, a fact that psychology has long recognized.

ANY EFFICIENT COMPUTATIONAL DEVICE REQUIRES goals in order to be able to decide which computations are good and which are less good or even bad. I've argued that goals can be indirectly defined by a collection of guidance signals. These goals can be quite abstract, maybe even too difficult to describe explicitly, but nonetheless very useful, like a word that is easy to use but hard to define explicitly. How can a goal signal be represented physically in an efficient computational device, whether the device is made out of biological parts or mechanical parts? The answer lies in the connection between goals and guidance—*valuation*.

In order to survive, mobile creatures must be able *to value* the resources available and the choices they can make to acquire those resources. An animal that moves in one direction forgoes resources it might have acquired going in another direction. For the same reason, even eye movements represent important choices. Seemingly small choices like eye movements or where to step next may seem irrelevant, but they add up vitally over time—for the good, or for the bad. Nature's economic realities have selected for creatures that can value their intended actions quickly and accurately. Value is the central concept here, and its quantitative study, in the context of goals and guidance signals, is a hot spot for theoretical work pursued under the name *reinforcement learning*. And it's here that the connection between goals, values, and guidance links measurable brain function to important mental function and dysfunction.

The ring of *reinforcement learning* could conjure disturbing images like those of Stanley Kubrick's *A Clockwork Orange,* where the lead sociopath, played by Malcolm McDowell, adorned with a hairnet of electrodes and props for his eyelids, is psychologically conditioned out of his antisocial behavior while Beethoven wafts along in the background. Fortunately, reinforcement learning is not the instrument of some futuristic dystopian culture, but is one essential way that animals and man-made machines can learn from experience. Reinforcement learning is not simply the stimulus-response learning familiar from introductory psychology, but is instead a quantitative framework for understanding how a machine can learn generally from experience. The field of reinforcement learning lies at the nexus of numerous fields, including computer science, control theory, and statistical learning theory. This is now such a large area that we will only require a small collection of the most basic concepts from reinforcement learning for the purposes of this book.

Let's start with stimulus-response learning in animals, the basic framework of behaviorism. In behaviorism, the central idea is to focus only on measurable relationships between input (stimulus) and output (response). In the twentieth century, behaviorists developed a kind of scientific philosophy around this perspective—all behavioral data would be explained only in terms of observable phenomena; hence the restricted emphasis on input and output. So what's the problem with that? What's wrong with behaviorism? Well, nothing, really, it's just woefully incomplete, and one can't really use it to explain cognition. Understanding what's going on in the brain (the black box for behaviorists) is impossible if all we observe is input-output relationships. Why? Because between the input and the output, there are innumerable brain states that do not see the light of day but have large effects on behavior. All adaptive creatures are vastly more than input-output boxes, and to understand the connection between brain and mind, one must take these unseen brain states seriously. This is even true for something as comparatively simple as the 1997 Mars rover called *Sojourner*, a six-wheeled vehicle designed to roam around the surface of Mars. After landing, a problem developed in its programming and Earth-based programmers had to upload new software to fix things. The new programming, encoded as a stream of ones and zeroes, was broadcast to the waiting rover on Mars, a broadcast with a delay of about fifteen to twenty minutes. During the upload, *Sojourner* basically just sat still and did nothing, just like a discharged Energizer Bunny. *Sojourner*'s brain was clearly being flexed internally but without any external symptoms of this activity.

A behaviorist account of this episode would not guess that all sorts of new internal instructions were being loaded and tested. While behaviorism provided rigor for a large part of psychology in the mid–twentieth century and still directs some learning experiments today, we now recognize that to understand brain function we must model and measure unseen brain states interposed between input and output. Reinforcement learning provides a natural mathematical setting for modeling these underlying brain states and relating them to observed behaviors. Reinforcement learning connects goals, guidance signals, and valuation in a family of mathematical models. We will consider our brain's reinforcement learning systems in detail in Chapter 3, but here let's set up the connection between goals, desires, and values.

A goal is a desired state. It usually changes with time (it's dynamic), and in the simplest cases, it's defined by important experiences like eating food, having sex, drinking water, running from predators, and so on. Many goal states can be defined simply by similar primary rewards. However, in humans, goals can be long-term and complex, like "I want to solve Fermat's last theorem" or "I want to be governor of my state." As I suggested earlier, these more complex goals don't have to be defined explicitly, but rather indirectly through a collection of guidance signals. Guidance signals are the error signals that tell the system how to adjust when deviations from the goal state occur. These deviations from the goal state provide a terrific model for the concept of *desires,* which inform the system how to adjust in order to move closer to achieving the goal state. Framed this way, goals and desires (guidance signals) become well-defined mathematical constructs.

A detective's goal could be to track a criminal's car in traffic without getting too close or falling too far behind. As his car deviates from this goal, his brain produces error signals that provide information for course corrections necessary to maintain the goal. This example shows that the goal signals are not static but generally change dynamically—the criminal's car is shifting lanes, turning at cross streets, and changing speeds. The detective's brain generates moment-to-moment course-correction error signals, and these are the mathematical rendition of desires (relative to the goal)—the drives that direct the system back toward its goals. In the human brain, these errors are generated as ongoing, dynamic differences in the value of the current state and the goal state. These values are numbers stored in the brain, and changes in value due to movement, internal thoughts (simulations), or surprising new information can generate the errors. In the language of reinforcement learning, stored values are like learned "wants." Our brains possess many sets of values tuned to specific behavioral and perceptual problems that we must solve to stay alive. These sets of numbers are called value functions, and it's changes in value functions that produce the guidance signals. The brain starts out life richly preequipped with lots of value functions, but these are just to get it started and to help frame difficult problems. As time goes by, the brain actually learns a vast amount of new values—that is, it learns what it should "want."

THE IDEA OF A VALUE function or valuation mechanism is not complicated. A value function tells a machine the value of the current state of affairs—simple as that. In the everyday world, where the machine is the brain, a value function does the same thing—it assigns a number to an outcome, whether or not the outcome is due to external experience ("my stock decision earned a positive return") or internal experience ("I feel sad that my job has changed for the worse"). In an efficient computational device, valuation is essential to help the machine decide how to allocate its resources, that is, its computing time and ultimately its limited energy. To be adaptive to new environments or simply exist in a highly variable environment, an efficient machine must have a way to update its prestored values and to learn new values altogether. The description above equating values to learned "wants" shows that the ability to learn to want something is a symptom of an efficient computational device.

There is now growing evidence from work in humans, monkeys, rodents, bees, and bacteria that valuation mechanisms can be correlated directly with physical measurements. In rodents, valuation mechanisms have been probed with food, sugar water, and sex. In bees, sugar water is the reward of choice. But in humans, valuation experiments have used an array of stimuli, including sugar, water, beautiful faces, art, diet soda, erotica, money, and so on. They have also been examined under conditions where the receipt of such rewarding stimuli is uncertain. Without belaboring the point, there is a wealth of biological evidence that creatures set goals and pursue them and that the idea of valuation functions is a clean way to try to capture goal-setting and goal-seeking behavior in mathematical models.

In Chapter 3, we introduce one particular goal-seeking system, the dopamine system, and show how it works to guide human and animal behavior. This system is vital to our survival and is known to be hijacked by drugs of abuse, damaged in Parkinson's disease, and malfunctioning in a range of psychiatric illnesses.

BRAINS OF ALL SORTS ARE amazing devices that carry out computations that no existing man-made machine can remotely match. Show me a computer that can control the movements of a housefly whizzing about and landing on the ceiling unharmed after avoiding two swats from my morning paper. Or a machine that can move like a wolf tracking prey for many miles through complex wooded terrain. Not to mention the linguistic feats of a child. These accomplishments become more amazing when we *divide by the energetic cost* of these computational wonders. For biological computers, the computational punch per unit cost is unrivaled. Human computer engineers still have a lot to learn from the long biological experiment that has taken place on this planet. Although a lot of work has focused on how to make machines learn from experience, much less effort has been directed at the same pursuit but adding the extra requirement of minimizing energy cost. In the real world, this stricture forced novel styles of computing and representation on biological computers, and we should expect to use some of these solutions in the next generation of our computing machines.

MY RABBIT
KNOWS WHAT TO DO

How Planning Makes Decisions in Advance

Model everything: This is
the second law of efficient computation.
ANONYMOUS

ACCORDING TO GARRISON KEILLOR the fictional town of Lake
Wobegon is a place "where all the women are strong, all the men are
good-looking, and all the children are above average." Nonfiction
mortals share this kind of self-deception. Humans don't like average,
and furthermore, we all like one side of average much better than the
other. Now, in the real world, average is actually very hard to hit. No
family produces exactly 1.8 children, and very few women are exactly
5 feet 3.7 inches tall. But near these averages and others like them
reside lots of folks, and some actually land on the wrong side of
average—below it. In fact, we all do. That's right, on some measure—
beauty, money, friends, digestion speed, or whatever—all of us will be
below average. But we almost never admit it, especially to ourselves
and only rarely to others. Why should we deceive ourselves?

The real question here is whether there is some kind of lower-
level reason, like computational efficiency, that explains the way we

model ourselves and why these models sometimes contain a bit of deception. Let's first consider the kinds of processes that our brains must model just to keep us alive. By adding our usual efficiency arguments we will generate concrete reasons to explain why our models of ourselves should necessarily contain a dose of deception. Before reaching that point, we must first discuss the blurred distinction between computing machines and algorithms.

HERE IS A REMARKABLE fact: The principles of efficient computing apply equally well to algorithms, not just to things that we usually consider to be physical devices. An algorithm is a procedure for accomplishing some task. For example, a step-by-step procedure for baking a cake, computing the volume of a cylindrical storage tank, or searching a list of items for one called "soccer ball." We don't usually think of an algorithm as a device, probably because the word *device* seems so physical and *algorithm* so abstract. However, any abstract sequence of steps is indeed a device, and some algorithms represent more efficient devices than others. If an algorithm can be thought of as a device, then it, too, must be subject to the same efficiency ideas outlined above. How should we measure the efficiency of algorithms?

Consider the problem of summing up the numbers in a long column in a financial spreadsheet. There is a simple algorithm for doing so—store the first number, add the stored number to the next number in the list, replace the stored number with the sum, and repeat until the list is exhausted. We could run this algorithm on a hundred thousand numbers using a modern personal computer and it might finish in less than ten milliseconds (ten thousandths of a second). Suppose instead that we ran this same algorithm on the brains of ten humans. Each human adds up a little section of the list and afterward all these results are added together by one of them. The total takes about a day to complete. Very inefficient, especially since it likely gives the wrong answer. This illustrates one way to measure the cost of a specific algorithm. Here, we keep the algorithm fixed, and measure the costs for running it on two different machines. We might do this across a whole collection of different computational devices to measure the average cost of running it, something we might call the *average execution cost* for this particular algorithm. If we knew this average, we could then decide whether a particular machine could run the algorithm at a cost below average. Big deal; this seems important only if we have a collection of different machines at our disposal and some important algorithms that will run on all of them. In the brain we do, but put this idea on hold briefly while I introduce a second idea about measuring the efficiency of algorithms.

There is a close cousin to the execution cost. Instead of averaging across machines for a given algorithm, let's keep the machine fixed and average across *equivalent algorithms,* that is, algorithms that produce the same answer. Our intuition suggests that algorithms can be analyzed abstractly without any reference to the machine on which they run; however, this intuition is wrong. There is always an implied machine in the background when analyzing the function of an algorithm, and it's better to be explicit about its assumed capabilities.

Why are the execution costs and intrinsic costs of algorithms important measures for an efficient computing system? The short answer: flexibility. A truly efficient computational device must be adaptive if it solves hard, real-world problems; it must be able to respond to unanticipated changes in the world around it and unanticipated errors that arise in its own operation. It needs not just one algorithm for any particular problem, but many algorithms that solve the same problem in different ways. To remain flexible, it must anticipate that there will be instances where it will need to get the same answer, but using different data, running under different conditions, or possibly using a dramatically different machine type. But how can a single computational system contain multiple machines? We must remember that machines can be simulations running on other machines. This is the central idea behind the virtual machine concept, an idea originally invented by Alan Turing in the early part of the twentieth century. Turing showed how one machine could run a program that simulated another machine, and he used the power of this idea to prove important properties of all algorithms no matter where they were running. This idea is familiar to anyone who has ever used an emulator program, a program that runs on one computer and simulates all the actions of another computer. The emulator simulation can be fed problems and data in the same way as the "real" device that it simulates. The idea of virtual machines blurs the distinction between a device and an algorithm. The emulator is nothing more than a complex algorithm for simulating another kind of computer.

STORYTELLERS ARE NOW EXPLORING this concept in movies like *The Matrix* (1999), written and directed by Andy and Larry Wachowski, and *Dark City* (1998), written by Alex Proyas (who also directed), Lem Dobbs, and David S. Goyer. These movies appeal to teenagers, but they also strike a strange chord in a lot of adults. How could we ever tell if we're just complex software running in some great simulation?

About ten years ago, my own daughters asked the same question. One project in my group was a simulation of bee decision-making using simple model neurons thought to influence learning and decision-making in honeybees. The simulations had graphical output that rendered each simulated bee on the screen. My daughters asked, "Where in the computer is that bee?" I explained that the particular bee on the screen was really a changing pattern of numbers inside the computer and that there was no possibility that this particular bee could get out. My youngest then asked, "But what if she wanted to?" I told her that since the bee did not know she was "trapped" inside the computer, she could not possibly want "to get out." I recall having to think carefully about my words before speaking (not my normal modus operandi). She persisted, this time with her now-classic hands-on-hips plus scowl-on-face pose. "Does she even know that she's just a bunch of numbers stuck inside a computer?" She sensed something unfair about the whole setup.

FINDING THE BEST MACHINE for a particular algorithm (minimum execution cost) or the best algorithm for a particular machine just states the same problem in different ways. In fact, you can see that the division between a machine that runs an algorithm and the algorithm itself can be somewhat artificial. It's algorithms all the way down—simulations of simulations and so on. In fact, it now seems very likely that our first-person point of view is itself a simulation running inside another simulation. Below, I present four quantitative properties that these simulations of self must possess to be consistent with our first-person experience. For now, let's shift to the real brain and see whether these general ideas take root there. In fact, they do take root right in the prefrontal cortex, that region of the cerebral cortex believed to be involved in rational thought, working memory, and decision-making. The prefrontal cortex collaborates with the striatum and the midbrain to produce a system that can generate one virtual machine after another chosen to solve some learning or categorization problem facing the system. In humans, destruction or damage to this region of the brain severely impairs the capacity to frame (categorize) correctly the problems that need to be solved and it dulls the afflicted individual's sensitivity to changing task demands. In some cases, subjects are highly distractible, moving from one problem to the next without any real progress. In other cases, subjects become stubborn, stuck using one approach to solve a problem even though it may not be working.

There has been an enormous amount of work in this area, but my point here is that the prefrontal-striatal-midbrain interactions form the best-known substrate for a system that searches for *simulated machines* best suited for a particular learning problem. Not just ways to frame or categorize the problem, not just ways to criticize actions taken, but an entire machine—an emulator that chooses an abstract space in which the problem is solved, runs parallel simulations of possible outcomes generated, and criticizes both using guidance signals similar to those sketched in the previous chapter. And the system doesn't stop there. It cycles from one simulated machine to the next hunting for the best solution, and it's likely that it is always running searches in order to anticipate the next machine that will be needed. The implication in these descriptions is that the brain can criticize an ongoing solution and switch machines when things aren't going so well. Lots of biological experiments support this view, though they use different language to describe the findings. My description is really a redescription of theories of the prefrontal cortex offered by others, but cast into the simulated-machine language that I'm using here. From our efficient computation perspective, this virtual machine search is much like dipping into a warehouse full of virtual machines, measuring each according to its predicted likelihood of working on the current problem, and choosing the best one to run. These machines aren't simply loaded into the prefrontal cortex, but instead concurrently run on multiple brain structures (prefrontal cortex, striatum, and midbrain). By thinking of this as a kind of "machine search," we can begin to build quantitative models of how such a search should be efficiently carried out and schemes for efficient storage of the virtual machines (e.g., putting similar machines near one another just like similar files in a file system). This same perspective has generated new ways of thinking about how our brains represent simulations of ourselves.

CHAPTER 1 ESTABLISHED THAT communication is expensive, especially communication over long wires, including the biological ones—hence the "save bandwidth" principle. We might therefore expect evolved brains to take dramatic steps to minimize communication. The single best way to accomplish this is to "stay off the lines," and this edict gives us one of the key insights for this chapter. In the world of cell phones, you can stay off the lines by knowing a lot about the person on the other end of the line and what she is likely to be doing.

Suppose that my wife and I are going to meet at the airport to catch a plane together, but we must travel separately because we are first going to pick up our children from different schools. The schools are on opposite sides of the city, but we start out from home together and we both have watches. We are synchronized at the start, carrying independent clocks, and we both have a model (in our heads) of one another—how long we should both take to get to each school and on to the airport. In this scenario, imagine that our cell phones are not well charged and can only support one or two brief phone calls to one another before expiring. If all goes smoothly, then I can anticipate roughly where my wife is in her journey to the airport and she likewise can estimate my position and likely time of arrival. Now suppose that I experience some traffic jam unexpected for this time of day. This event illustrates why it's important for me to model my own journey *and* my wife's journey. I can query my internal model of my own progress to decide whether this delay will make me significantly later than *she* would expect me to arrive—remember my model of myself is the same as hers. If the expected "error" between my actual expected arrival time and what she expects is too large, I can call her to send her a corrective update, "I'm going to be fifteen minutes later than expected." She simply updates her model of my progress accordingly, and we have kept the communication to a minimum by talking only when necessary, that is, when there was an error between our expectations and our actual experience. An unexpected event on her end can be handled in a similar fashion. By modeling our own and each other's progress, we "stay off the lines" and save communication time that our nearly empty cell-phone batteries have made precious. There are two morals to this story. First, to be efficient, my wife and I must carry two models: one for self and one for the other. Second, again for efficiency, the communications should mainly be about significant errors.

All complex computing machines share the same issues with me and my wife. They have different modules that must communicate with each other. Communication channels in any device are the single most costly item in terms of their consumption of space and time resources. To minimize the number of communication channels running between modules, they must be as self-sufficient as possible. The solution? Model the modules. Equip each module with a model of the operations of itself and all the other modules. This allows each module to simulate "where everybody else is" in their respective computations and "where everybody thinks I am." Taken together, this knowledge allows the overall machine to reduce its internal communications and hence operate efficiently. We should therefore expect those brain regions that are reciprocally connected to possess the ability to model one another. We might call this kind of feature a *mutual modeling principle* for any parts that were connected by communication channels. In a parallel computer with many processors that communicate with one another, this mutual modeling would allow interconnected processors to minimize communications required between them. When working perfectly, such a system would synchronize processors to start things rolling, and communicate intermittently only when unexpected errors arose within one of the communicating modules or when a reset was required. Of course, each processor is running dynamical models of itself and all of the others so that it knows exactly when it should communicate errors. This same reasoning can be applied to a range of human interactions by treating them as systems that must model one another in order to cooperate and compete.

So the best energy-saving brain is one that has no need for any internal long-range communication. In the real world this cannot be achieved, because the world is just too variable. But the efficiency principle for the brain is clear—separate brain systems that must talk to one another should keep models of themselves and all those to whom they speak. This is similar to a preferred cell-phone list. Consider all the speed dial numbers in your cell-phone address book. You have a fairly rich internal model of each individual or business in this list because you interact with them frequently. Otherwise, you would not waste a speed dial number on them. By analogy, the parts of an efficient computing machine that talk to one another most frequently should, by the principles of efficient computation, possess models of all their talking buddies, and their talking buddies are those subsystems to which they are connected. These models allow these separate systems in the brain to operate as autonomously as possible in order to minimize communication. It's in exactly this sense that we live in a kind of virtual world synthesized from the vast sets of internal models running on our brains, all talking to one another, but no more than absolutely necessary.

Internal models rely heavily on memory resources, but fortunately, memory is cheap. In fact, thanks to a thermodynamic analysis by Rolf Landauer and its later improvement and refinement by Charles Bennett, we now know that information storage using any physical medium can be made arbitrarily cheap, if one is willing to write the information down slowly. It's the erasure step that generates wasted heat. Bennett addressed these issues using theoretical arguments, but also by considering concretely the cost of writing and erasing using the real enzymes that read and write the ultimate memory tape—DNA. The work of these two physicists and others leads to the suggestion that the brain leans heavily on memory-intensive solutions to modeling, and that it would only erase information when absolutely necessary. This means that, like junk DNA, we probably have a lot of junk memory just hanging around in our brains (I know I do)—information that remains unused because it's cheaper to keep it stored than to generate wasted heat erasing it.

ANY CREATURE THAT PRODUCES rapid ballistic movements of any sort must be able to model the future quite accurately. Consider a rapid eye movement that takes the eye from looking in one direction to looking in another. The movement is really too fast for feedback signals to be of much use, so the neural circuits that control this important action simply produce a copy of where they are planning to send the eyeball, and this copy is distributed to lots of other neural systems. The brain knows quite well how to execute this movement. Its internal model of the eyeball and its dynamics is accurate enough to permit it to plan and execute a movement without any significant feedback, which in this case is too slow to be of much use anyway. This kind of simulation is called a forward model, since it's equivalent to simulating things forward into the near future. Many systems in your brain are known to generate forward models, and I suspect that the list will grow to include all significant brain systems in the future. A forward-modeling capability allows for all sorts of rapid computations, but it also allows all the other systems to anticipate and prepare for what's about to happen: The eye will move, the tongue will flicker out, the hand slap at a bug, and so on. The brain must be able to model all sorts of important issues about the near future. These simulation results would read like the "mother of all worriers." *Where will my eyes be next? Where are my hands? Where should each foot step next? Will I need more oxygen soon? Should I speed up or slow down the heart?* and on and on. Enormous amounts of monitoring and simulating are required just to stand up, breathe normally, and keep the heart beating at the appropriate rate. It's a wonder that our conscious experience doesn't really mess up our moment-to-moment existence, but thankfully, we have no need to consciously deal with all these measurements and models.

The capacity to model movements "forward in time" is presently a very active area of research in machine learning, robotics, and neuroscience. In all these areas, the problem of controlling movements is now known to be theoretically unsolvable in the absence of rich internal models of the parts to be moved and their predicted trajectories. This same conclusion applies to limb movement, facial movement (expressions), tongue movements, and the small constellations of movements required to reach, scratch, wave, and so on. These models are complex and poorly understood in detail, but we know that they exist. The lack of humanoid robots or humanlike movements in practical robots has not been for lack of attempts, but because the problems are just hard to solve.

The need for simulation even in the absence of feedback is captured quite well by the words of theoretical biologist William Calvin.

Plan-ahead becomes necessary for those movements which are over-and-done in less time than it takes for the feedback loop to operate. Natural selection for one of the ballistic movements (hammering, clubbing, and throwing) could evolve a plan-ahead serial buffer for hand-arm commands that would benefit the other ballistic movements as well. This same circuitry may also sequence other muscles (children learning handwriting often screw up their faces and tongues) and so novel oral-facial sequences may also benefit (as might kicking and dancing). . . .

Calvin makes an additional important point—redeployment. He suggests that the capacity for one ballistic, feedback-free movement sequence could have been redeployed to generate other, potentially useful, movement sequences. Calvin suggests here that mechanisms that permit fine motor control for writing might have also been used to sequence facial expressions and therefore slowly been usurped to also support this role. Whether or not we believe Calvin's prescription for how our simulation capacities evolved, the idea is clear enough. Once the brain could model sequences of action forward in time quickly and accurately, there was a capacity to represent what "could have been." And once this ability could be carried out "offline," the brain was off to the races. Simulating "what could have been" is a capacity central to learning, critical for representing ourselves and others, and it's an efficient way to know how and when to share information.

The core feature of any simulation is to emulate what "could have been" and learn from it. There's nothing particularly mysterious or deep about this idea. The brain simulates possible future scenarios, values the fictive outcomes of each scenario, and uses the valuation to help choose a course of action. Simulations that probe possibilities farther into the future give more information than simulations that do not, but they also cost more in energy. The capacity to run many simulations in parallel should yield better decisions than running only one or a few simulations at once, but here, too, the costs are higher. From the efficiency principles in the last chapter, we should expect that our brain's simulations will attempt to reduce the number of possibilities that it must consider and use a degree of parallelism that suits the problem at hand. These ideas are consistent with the reality that parallelism doesn't really scale very well. The payoff in usable computational speed per unit cost grows only mildly as the degree of parallelism increases. This reality may suggest why the massively parallel general-purpose computers of the late 1980s and early 1990s did not find a large, receptive market. Simulations of experience must also be equipped with some kind of critic signal to enable them to learn from the fictive experience that they emulate. Therefore, they must have fictive learning signals associated with them.

Why should our nervous system bother learning about things that didn't happen and choices not made? Shouldn't it be spending most of its learning on real experiences? Our nervous system indeed spends a lot of its resources learning from its own direct experience. However, it's possible, and cheaper, to also learn from others' experience or even one's own experience that reasonably might have happened.

Let's do a thought experiment. Imagine a small field with three patches of flowers in it. Each patch is labeled by the color of its flowers, one red patch and two green patches. Two bees are released in the field and each possesses an internal model of the value of colored flower patches. Bee X thinks that green patches deliver on average ten units of nectar and red patches are five times worse, delivering only two units of nectar. Bee Y has the exact same internal model. The bees are released into the field and Bee X flies to green patch #3 and Bee Y flies to the other green patch (patch #2). So far, so good. Bee X harvests ten units of nectar in her green patch, exactly what she'd expected. She flies back to the hive to report her harvest to the other bees. Her action does not generate any errors with her expectation, and her model of green patches continues to be ten units of nectar. Bee Y harvests one hundred units of nectar from the other green patch, ten times as much as she'd expected!

Bee Y is "happy" and updates her model of what green patches are likely to yield; in this case, she might increase this expectation from ten units to twenty units—her nervous system doesn't and shouldn't completely believe that a hundred units are going to be delivered from now on. What about the model of Bee X? If she never encounters Bee Y, then her model should stay at ten units. This is a rational use of her direct experience. But once Bee X learns about the other green patch from Bee Y, her nervous system will compute a regret signal, which is substantial here, and use this to update her model of the nectar to expect from green patches. She has benefited from the knowledge of Bee Y's efforts without having to experience directly the other patch. Her model of the world has been improved by using Bee Y's experience and assuming that things would have been the same for her. This is the efficiency of combining "what could have been" with actual experience to build a better model of the world. In this example, the bee uses real and counterfactual experience to build a model of nectar distributions in the field. Likewise, Bee Y should use the information from Bee X about ten units harvested from the other green patch. By sharing information like this and by possessing neural mechanisms that sense and respond to "what might have been had I sampled the other green patch," these two bees build a better model of green flower patches (in terms of nectar) and can make better decisions about them in the future. This is a contrived example that's not quite accurate, but it makes the point—since acquisition of resources requires learning good models of the world, an organism should use information about what might have happened to improve its internal model. The model is essentially a proxy for future resources. Better models mean better future returns. And that's a strong incentive for an organism to use and respond to fictive outcomes. In humans, imaging experiments have visualized the action of learning signals generated by real experience and fictive experience. It's important to understand that the learning signals derived from fictive experience are different from the emotional signals that may also result from fictive experience. An emotion such as regret is different from the covert, nonfeeling fictive learning signal that underlies all the pain and power of the emotion.

As URSULA K. LE GUIN said, "The only thing that makes life possible is permanent, intolerable uncertainty; not knowing what comes next." Uncertainty is what humans love and hate about the world. We hate uncertainty because it can deliver losses. We love uncertainty because it introduces the "maybe" factor—maybe I will win next time, maybe things aren't as bad as they seem, and so on. The maybe factor lures our decision-making machinery toward the future. Uncertainty long ago put evolutionary pressure on learning mechanisms, and creatures that could effectively reduce their uncertainty through learning survived better than those that couldn't. So uncertainty was the evolutionary question, and learning (model-building) was the answer. The problem with this characterization is that all uncertainty is not created equal. Some sources of uncertainty arise simply from ignorance and can be reduced through learning, while other sources of uncertainty are totally unaffected by learning. This latter kind of uncertainty is called *irreducible* since it cannot be reduced through experience and learning. This idea is familiar to all of us, but it's just hidden a little bit by the lingo. Suppose that we put ninety red blocks and ten white blocks in a basket and shake to mix them up. Now imagine that you start to sample blocks from the basket, but you begin with the conviction that there are an equal number of white and red blocks so you believe that there's a fifty-fifty chance of getting red. This model of the basket is wrong, but you don't know this at first.

Now suppose that you sample blocks in groups of one hundred, but replace each individual block after it is sampled. In each group, you record the number of red blocks. After repeating this procedure a number of times, you will soon notice that the average number of red blocks per hundred is higher than you'd expected and the variability in the number is less than you'd expected. Basically, your experience with the basket deviates significantly from your prior expectations about the basket, and so you decide to adjust your beliefs about the basket. You shift from a fifty-fifty expectation of red to white to something close to ninety-ten. You have built a better model of the basket's true contents by exploring the basket, experiencing a lot of variability, and using the outcome to adjust your expectations about the ratio of red blocks to white blocks. The initial unexpected variability has been reduced by your sampling and model adjustment. The initial deviations from your expectations (your prior model) resulted from your simple ignorance of the true ratio of red to white blocks. You have reduced your ignorance of the likelihood of selecting red and the associated variability in this likelihood. This is the *reducible uncertainty,* that is, uncertainty that you have eliminated through exploration and learning. But once you settle on about ninety percent for the likelihood of getting a red, that's it. There is nothing more to learn. There will still be variability from sample to sample; however, no amount of learning can reduce this variability, since it's intrinsic to the problem. You will get sequences of selections where the percentage of red blocks is above or below ninety percent. This variability is *irreducible uncertainty.*

Let's import the efficiency arguments from before to draw an important conclusion about uncertainty detection. In a novel setting, you will encounter a lot of uncertainty, which creates two crucial mental tasks on which your survival depends. The first task is to take actions to "learn away" the reducible uncertainty in this new situation. The second task is to know when to stop. Without a finely tuned "you've learned all you can, now stop" signal, your brain may encourage you to keep attacking a problem in which there is a lot of irreducible uncertainty—an encouragement that breaks every efficiency rule that we have discussed and will waste an enormous amount of energy. Efficiency mandates the existence of uncertainty-detection mechanisms in the brain, especially mechanisms that can quickly tell the difference between reducible and irreducible uncertainty. Uncertainty detection and reduction is an active research area in theoretical neuroscience and computer science, and we can now identify a neural system involved in these essential functions.

WE ONCE HAD A black rabbit named Roger who taught me two important facts about rabbits. The first fact is that rabbits can growl. That's right, the typical *grrrrr* that we expect from a dog can also be emitted by rabbits, or least by our Roger. The second fact that I learned from Roger was that he practiced escaping. Inside the house, Roger would run specific routes around the furniture and through rooms. And these routes were run at full rabbit speed! OK, not so remarkable. "Rabbit runs at full speed through house, film at eleven." The remarkable feature was that Roger would run exactly the same route over and over again. I once watched him run an exact route six times without any source of provocation. Each time he would finish,

he would just sit still for a bit, then hop back into the kitchen and begin again. Roger did the same thing once we let him explore our
fenced-in backyard. He would plot out some course through bushes,
behind trees, and over small open expanses of grass. At first, he would
sort of hop along this path. Later, he would run the route at full speed,
making the same sharp turns at the same places in the path. Maybe
Roger had some kind of mental affliction, but his behavior was very
suggestive to me and makes a point for our discussion.

While observing Roger, I thought that his route-running might
well be a good strategy for escape. By his practicing a seemingly erratic path repeatedly, Roger's nervous system would not have to take
those extra moments to decide to move right or left; instead, Roger
could execute a practiced, but effectively erratic, path at slightly higher
speeds. Roger's path would be harder for a predator to follow, since
the predator would have to detect Roger's next move, make a decision
to change direction, and hope that Roger has not "juked" him. This
facile description of Roger's strategy overlooks something familiar to
all poker players. Since Roger's brain has overplanned his route, his
next move may reveal itself in small muscle twitches or limb movements that anticipate the upcoming "erratic" turn. In poker, this is
called a tell—a telltale sign of an impending act reflected somehow
in an opponent's movements, sounds, or body language. Although
Roger gains valuable speed by planning and practicing his escape
route, he loses on occasion because speed is gained at the expense of
revealing Roger's next likely move. Roger's escape plan needs some
spice, some real unpredictability without being so unpredictable that
the route does not lead to an effective escape. What can Roger do?

Roger needs to add a source of irreducible uncertainty to his escape algorithm. He needs a fraction of his moves to be erratic to his

pursuer and to himself. Neither Roger nor his pursuers can use learn-ing to get rid of irreducible uncertainty. If Roger doesn't know with certainty his exact next move, then he can't possibly betray it to his pursuer by providing anticipating information through subtle signs in his behavior. This information-processing strategy of adding irre-ducible uncertainty will slow the predator, since it cannot in principle know exactly when and where Roger will make his next move. A clever pursuer might recognize that Roger's escape algorithm included irre-ducible uncertainty and attempt to model how often Roger uses it. This would help the pursuer, but it could never solve the problem ex-actly. In his book, *Decisions, Uncertainty, and the Brain,* neurophysi-ologist Paul Glimcher argues that all biological decision-makers need a source of irreducible uncertainty, but he starts with a different set of assumptions about the economics of decision-making in biological creatures. Whether one takes a learning perspective (as I am here) or an economics perspective (as Glimcher does), one consistent predic-tion emerges—the brain must possess something akin to a random-number generator.

Roger can make his escapes hard to predict by adding enough ir-reducible uncertainty to make his moves unpredictable, but not so much that his escape is not effective. Roger's reward for success is to stay alive and the predator's reward is a nutritious meal. By deceiving himself a little bit, Roger makes himself a more difficult prey to catch. True self-deception requires that Roger's brain have a source of ran-domness that cannot be modeled by Roger's brain or his competitor's brain. If Roger's escape algorithm was merely complex, over time small telltale signs might bleed into his observable behavior, and his pursuers might learn the rules that generated his escape tricks. How does Roger's tale impinge on human social exchange?

Early humans had to solve many of the same survival problems that rabbits do. Although rabbits are very social animals, the social needs and interactions for rabbits do not match those expressed by humans. Humans have strong social instincts that draw them close to one another and allow them to cooperate in novel ways. These instincts are subtle, and form the foundation of our complex, multilayered societies; therefore, I won't deal with these issues in any depth. I want to make one simple point about the presence of irreducible uncertainty in our neural representations of ourselves. I want to indicate why we should expect our mental representations of ourselves to possess a calculable amount of true uncertainty, that is, true self-deception. In the simplest social exchanges, two parties trade something for something. Let's try to imagine a social transaction as an efficient computation. What ingredient could allow both parties to maximize their yield, regardless of what is exchanged? There must be mechanisms that prevent one party from always winning at the expense of the other, since this would eventually erase the losers. In species other than humans, this issue is studied under the title of reciprocal altruism. For example, this issue has been addressed at length in numerous brilliant papers on social insects, but there the problem is different because the willingness to win or lose at another's expense can be traced to shared genetic futures. With my fellow humans, the problem of reciprocal altruism, or reciprocal anything, is different. My speculation here is that social trade instincts in humans are best encouraged by building irreducible uncertainty into my internal simulation of myself.

During a social transaction with another human, I must model myself and my trading partner. My suggestion here is that my simulation of myself in the trade must possess some irreducible uncertainty that neither my brain nor my trading partner's brain can model. This will insulate both of us from total exploitation by the other. And to make this mechanism efficient, I should advertise clearly to my trading partner that I indeed use internal randomness in my trading algorithm; that is, there is a bit of an erratic component in the way that I will trade. Without this advertisement, my partner may expend a lot of energy trying to learn my strategy—more energy than is necessary, since some part of my strategy, the irreducible part, is not learnable. For the social exchange to be efficient, I should save my partner this effort and simply advertise my irreducible uncertainty. OK, one side of the trade puts irreducible uncertainty into her exchange strategy and advertises this fact. Why would someone trade with this person? He wouldn't. For this trick to work, my trading partner must be willing to put up with this element of risk in my behavior. And this should not happen unless the exchange is balanced just like a chemical reaction. I'm willing to do the exchange if the potential loss from my partner's uncertain behavior is equaled by my expected gain from the exchange. And this need is symmetric—both partners need this to be true. How could this ever happen? Trust—a mechanism that induces someone to carry out a social exchange despite the risk of loss. And there must be mechanisms in both brains that create the perception that the risk in social exchanges is mitigated; otherwise, the two parties could just find other nonrisky trading partners. But truly nonrisky trading partners can eventually be exploited so they would be quickly eliminated from a population.

THE ABILITY TO simulate oneself and the environment into the near future is a capacity crucial for selecting good sequences of actions, that is, actions that do lead to food, sex, and safety and don't lead to hunger, frustration, and harm. Let's consider an example. Suppose that I was hungry and saw a fruit tree on the ledge of a rock face. It's quite high up and I must climb up the rock face just to get to the tree. Also, at the level of the higher tree branches is another ledge on which I might lie down to reach out to some of the higher, fruit-laden branches. To decide whether to attempt the climb up to the tree's ledge, I must be able to mentally translate my body up to the ledge and imagine how close to the edge I will get when I reach or jump slightly for the fruit hanging off the lower branches of the tree. I must also be able to estimate the short- and long-term value of actually acquiring the fruit. These simulation capacities help me decide whether such a reach or jump is worth the associated risk. Try to imagine these scenarios yourself. If I drew you a depiction of the tree and the two ledges on a piece of paper, it would be an even easier mental task. We can all use our imagination to *scale* our bodies to fit the relative scales depicted in the drawing—that is, the relative sizes of the tree, rocks, height of the ledge, and us. We can also *translate* our scaled selves to an imagined position on the lower ledge housing the tree. It's also quite intuitive to imagine which fruit is the simplest to reach from this ledge—the ones hanging off the lower branches. From the higher ledge, fruit on the higher branches is easy to reach, but the higher ledge may seem a little less safe. Now imagine that you are lying prostrate on the top ledge that is level with the tree's higher branches. Would you be willing to reach out and grab the fruit? The answer depends on your penchant for risk-taking and the estimated value of the fruit. There is also a bush on the lower ledge next to the tree; it's large enough to conceal your entire body. Your mind can imagine waiting behind the bush for five minutes, and later, the same "you" emerges to decide whether to reach for fruit. The overall point here is that it's very easy for you to simulate (imagine) these transformations, and during the exercise, the "you" involved (the point of view) doesn't change; it's invariant. *Invariant* literally means "unchanging." What is unchanged? Your simulated point of view. When I asked you to imagine the movements and actions, you didn't suddenly forget who you were or feel dramatically different. Something very important remained the same. At least this is what your perceptual machinery presents to your first-person perspective. In these scenarios, there is nothing particularly special about simulating a version of oneself and translating it to the first ledge, translating and rotating it to be on the second ledge, or letting it wait behind the bush for some time. A stepladder could do just as well. You can substitute the image of a small stepladder with three rungs on it, carry out the same simulations, and answer questions about the disposition of the ladder after moving it to the various locations.

The capacity to imagine all sorts of would-be scenarios through simulations of experience confers a distinct advantage for decision-making and planning. For example, "If I am standing next to the tree I can safely reach the fruit," or "If I am chased, I could climb the rock face, hide behind the bush for an hour, and reemerge later unharmed." Notice that there is one left-out component here, the implied *were*. All these simulations are conditionals. "Were I to lie down on the top ledge, the fruit would be about X inches away from my hand," "Were I to hide behind the bush . . . ," and so on. These simulations are counterfactual experiences that have not yet happened. They are not real experiences, but they are like real experiences, and they influence our real actions and they take up valuable processing time in our brains. And we shouldn't be too surprised that our nervous system possesses this capacity. Otherwise, how could we account for the emergence and use of the subjunctive verb forms? I'm no linguist, but it seems reasonable that this capacity was around long before hominids could use words. Being able to run a bunch of conditionals like this is essential for all mobile adaptive creatures to make good decisions, not just creatures as complicated or talkative as humans. It's also clear that humans can do these simulations over their possible past experiences as well as their possible future experiences.

Do these simulation capacities reveal anything else about mental function? In fact, they reveal something very important about "mental objects." Psychologists Roger Shepard and Stephen Kosslyn carried out a series of elegant perceptual and neuroimaging experiments to probe visual imagery capacities similar to the thought experiments discussed above. However, here I want to emphasize a completely different aspect of these capacities. The mental capacity to translate, rotate, and scale imagined objects (including people and points of view) is part of the tool kit that makes up our capacity for doing intuitive physics, and it most likely is present in all mobile animals to some degree. This tool kit also includes our capacity for object permanence; that is, the capacity to maintain a mental representation of some object or idea, invariant through time. The classic example involves a baby, a sofa, and a ball. Depending on the age of the baby, this is a formula for about twenty-two seconds of fun before something else entertaining needs to be found for the baby to do. Roll the ball behind the couch. Young babies do not understand that the object that is "out of sight for some time" is still there but simply obscured from view. Their internal model of objects does not include a functional equivalent to "persists through time." An adult finds this almost absurd and the average adult "knows" effortlessly that the ball now will be the same as the ball in five minutes. At some point, usually around eight months or so, this capacity comes online in the developing infant. In the simplest terms, the baby is now able to model parts of the world as persisting through time. The famed psychologist Jean Piaget first described object permanence, but mainly in terms of the capacity to represent an object not seen; however, here we focus on the fact that this requires the capacity to represent the object as having *persistence through time.* That is, objects are those things that are invariant through translations in time.

What are we describing here? What do all these transformations tell us about our nervous systems and our mental software? They tell us about one important feature of how our nervous system represents our conscious point of view. This point of view is invariant with respect to three important transformations: (1) translations through space, (2) rotations through space (e.g., change view to top of overhang), and (3) translations through time (persistence through time). It's the same object before and after these transformations. The "object" that is unchanging here is the "you," or rather your nervous system's representation of your first- and third-person point of view. And these capabilities are not limited to models of just you, but extend to any objects. You can easily simulate a bucket undergoing all the same transformations outlined, after which you could answer questions about the disposition of the bucket. In addition to physical transformations, people are what we're really good at simulating. We have processors to emulate the internal feelings of other people. This is our capacity for empathy. It's a hot area for neuroimaging studies, and recent work by Tania Singer, Chris Frith, Ben Seymour, and Ray Dolan has begun to pave a new neural understanding of empathy. I don't know whether Einstein was good at simulating feelings, but he had an uncanny ability to map complex physical problems onto his internal capacity to simulate physical transformations. This is how he made complex physical ideas so accessible to a wide range of scientists; he mapped the problems onto simulation capacities present in the minds of a wide audience.

The conclusion: Our nervous systems can generate models of ourselves and objects that do not change relative to translations through space, rotations in space, and translations in time (persistence through time). These models can be the first-person perspective or third-person perspective; both survive these transformations. This thought experiment shows us a set of properties that mental models of self must possess. A mental model of self also has other capacities, but I'm just focusing on these transformations here. Now the remarkable part—objects that are invariant in just this way are exactly the kinds of objects modeled by physics.

PHYSICISTS LONG AGO decided that their models of "physical objects" should not depend on the point of view of the modeler (observer). The physical description of these objects should be invariant to translations in space, translations in time, and rotations in space. These are good requirements and we know how to represent them mathematically. Imagine a physical model where a coffee cup with a handle on the right was considered to be different from the exact same coffee cup rotated around so that the handle was on the left. Our mind does not label these "views" of the cup different objects. And this feature is efficient for two reasons: It saves on the number of variables required and it allows us to share experiences with one another. Instead of modeling each view separately and considering each perspective as a separate object, it would be more efficient to represent the cup with handle in a way that was invariant to rotations of the observer's point of view. This same reasoning applies to persistence through time, and invariance to spatial translations. It would have been inefficient for our brains to represent objects as different just because they were translated to another position or had been sitting around for five minutes with no appreciable change. Now, of course objects do change through time, but they only do so by gaining or losing energy. If nothing dramatic happens to a cup, then my neural representation of the cup now and ten minutes from now should be the same. It's an economical choice of representation.

In psychology, what is less well appreciated about these invariances is that they are an equivalent way of describing the conservation of energy (invariance to time translations), conservation of linear momentum (invariance to space translations), and conservation of angular momentum (invariance to rotations in space). Yep, the three great conservation principles of physics! The whole of classical physics can be derived using these principles. However, given our previous discussion, are we really that surprised that the needs of our physical models matches the invariances that our nervous system first required of their internal models? Remember, the ability to simulate translations, rotations, and persistence through time long predates our explicit theories of physics. Lots of animals without language and without schools can simulate experience in just the way we have described. From an evolutionary perspective, the "objects" of perceptual experience (conscious or unconscious) were invariant in the way I describe long before we discovered how to capture our experience in sets of mathematical equations (physical models) and transmit them through culture.

Why is this realization about invariances important? One reason is that it prescribes key mathematical properties that internal models of self must possess. Such an insight applies to both first- and third-person models of our internal "point of view"; they must be invariant to translations in space, translations in time, and rotations in space. Any informational object in your nervous system that might represent a personal point of view must at least possess these invariances. We now have some nice mathematical properties that simulations of self must possess. Lots of mathematical machinery is already built up to describe objects in a manner insensitive to these invariances. These descriptions are called coordinate-free descriptions by physicists, but for our purposes they suggest one way that humans can "share" experiences with one another and understand the intentions of others. The answer is that humans appear to come preequipped with a way to represent someone else's point of view that does not depend on that person's idiosyncratic "coordinate system" for representing himself. This may have been a trick stolen from the visual system. The idea has been floating around for years that the visual system uses coordinate-free representations of visual objects. I'm simply taking one more step to get a glimpse of how we can know what someone is thinking even though his brain may use completely different coordinate axes for representing the data in question. In fact, it may be that our thoughts are in general expressed in a coordinate-free internal language.

The observations about invariances also raise an important but unanswered question concerning the counterfactuals, the "what if" simulations described above. Based on a set of possible simulated futures, which one should my brain choose? Each counterfactual future is a possible time line for my experience. Is there a principle by which my brain should combine these fictive trajectories of future experience in order to pick the "best" one? No one knows how to answer this question yet. In decision sciences, each possible time line is usually represented as separate from the rest and possessing a value to the decision-maker that can be captured by a number. However, this approach tends to focus mainly on how to assign value to each time line rather than on how the time line should be represented mathematically and interact with other possible time lines. I suspect that these possible futures interact somehow before one is chosen. One speculation, hinted at by the work of physicist David Deutsch, is that one should let all the plausible counterfactual simulations "bump into one another" to produce the best average outcome. This is an area under theoretical investigation today and may be the tip of a coming fusion between psychology and physics.

4

SHARKS DON'T GO ON HUNGER STRIKES

And Why We Can

Some goals are more efficient than others;
this is the third law of efficient computation.
ANONYMOUS

SHARKS DON'T GO on hunger strikes. If they are hungry and food is available, they eat. No remorse. No angst. Not much thought before or after. Sharks don't worry over the mental status of their prey, nor do they fret over how the death of a fish or a seal or even my neighbor will affect their image of themselves. Their behavior is easy to conceptualize as instinctual, so we really don't blame them for killing fish. How about dogs? If a dog is hungry and there is dog food around, the food will be eaten. However, there are exceptions; dogs aren't just sharks with four legs, a tail, and bad breath. We all know finicky dogs who have taught their owners to give them certain kinds of food, or else. I fall into this unfortunate category. Zealous dog owners routinely fall prey to the newest pet food craze—especially if it promises some new twist on dog health. If the newest "healthy kelp extract" food is presented to your dog, he may pout in some unique way or simply refuse to eat. Dogs can inhibit their need to eat, but not forever. And I know of no dog that can refuse food to the point of death. But humans can. Humans can deny even their most basic survival instincts for seemingly arbitrary ideas. Humans routinely starve themselves to death because of some perturbed mental image of their body or some political end. Humans can go on hunger strikes. But such behavior can get even crazier than mere hunger strikes.

Just consider the 1997 mass suicide by the Heaven's Gate cult in a mansion in exclusive Rancho Sante Fe, California. Thirty-nine adults between the ages of twenty-six and seventy-two committed mass suicide based on a shared belief that there was a spaceship hidden behind the tail of the comet Hale-Bopp waiting to take the group to the "next level." Their leader, Marshall Applewhite, had been surgically castrated, as had numerous male members of the group, apparently in preparation for the "next level." Pretty crazy stuff—and very painful for the living family members left behind. Hale-Bopp was indeed an inspirational sight, and comets have long inspired awe and image-making in humans. However, most of us aren't willing to bet our lives that a spaceship is slinking behind the impressive tail of some comet with the "next level" as its final destination—a fact contradicted by evidence from our most powerful telescopes and the sheer improbability of the claim. Why is this story so disturbing? I think it's because we simply cannot make up a rational tale where these acts make sense. It doesn't fit any familiar story of human behavior, and its inexplicability adds to the shock of the act. I can't lift the veil on these events, but this tragedy highlights an important power possessed by all our nervous systems: a superpower.

The amazing part of the Heaven's Gate story is that the cult members used an abstract idea—going to the "next level"—to veto their powerful instincts to survive. This act defines a behavioral superpower—the capacity to veto survival instincts to the point of death. Sharks cannot do this. Dogs can wait awhile before eating, temporarily inhibiting their desire to act on their hunger; however, they will not starve themselves to death just to make a point. Let's be sure to distinguish this idea from self-sacrificing actions based on a kind of genetic balance sheet. Dogs will put themselves in danger (and may consequently die) in order to protect their young, temporarily suspending their instincts to stay alive. Certain species of ants will literally explode as a defense mechanism, their deaths aiding the continued survival of their brethren. Biology provides a long list of these ultimate sacrificial acts for relatives, but there is a catch. These behaviors operate with genetic imperatives behind them, and reflect a kind of survival balance sheet: trading one life now for a new life (with related genes) in the next generation. Instinct denial for an idea is something entirely different, and almost ironic: a survival machine with the capacity to choose to die, not just for a genetic reason, but for an arbitrary idea.

How can an idea gain that much behavioral punch? Unless there's something magic at work here, we are forced to conclude that the idea of the spaceship and its ultimate purpose insinuated itself into the devotees' nervous systems, commandeered their behavior, and caused them to take their own lives. A mere idea hijacked the controls of these people's brains and drove their bodies off a cliff. To do this, the ideas about the spaceship and the "next level" had to represent something tremendously valuable to the believers' nervous systems— something more valuable than life itself. Modern neuroscience has begun to sketch broad-stroke answers to how such a superpower works, why it goes awry in drug addiction and mental illnesses, and how, despite its obvious dangers, it equips us with a powerful tool for cognitive innovation.

In the last chapter, we discussed the relationship between rewards and goals, and this gives us the basic idea behind this superpower. The abstract idea of "going to the next level" defines a goal to the nervous system by acting as a reward signal. It's easy to see how food and water can act as rewards; there is a direct connection between them and ongoing survival. For the cult members, "going to the next level" acts with the power of food, water, or sex and gains the attention and respect of the rest of the nervous system, which tries desperately to acquire this abstract goal.

With abstract goals acting like food and sex, it's no wonder that the guidance signals that chase the goal may produce chains of bizarre behavior, including the macabre series of steps carried out by the Heaven's Gate cult members. What do we mean by *rearrange*? We mean "adapt." The nervous system adapts and adjusts its internal structure to try to acquire the goal. Neurons adjust, the connections between neurons adjust, and even other cell types in the nervous system adjust—all under the influence of the guidance signals. This is multilevel learning: The nervous system's parts and their interactions learn (adjust), guided by the new goal. As the internal neural structures adjust to achieve the goal, cognition literally changes, and often in hard-to-reverse ways. A familiar example is drug addiction.

Anyone who has ever dealt closely with a drug addict knows that she no longer views the world the same way. One reason is that drugs of abuse derail exactly the guidance signals that we have been considering. Under the perturbed guidance of drugs, an addict's nervous system readjusts complex goal-seeking mechanisms and changes her mental world; it's very much like a pathological software update. Unlike those in your personal computer, software updates in the brain change neural structure, that is, they rewire you. A much better understanding of these software changes in the context of our detailed understanding of drugs and their effects is required before we achieve real cures in this domain. There is a growing effort for computer scientists and neuroscientists to weigh in together on just these issues. For now, let's stay at a high altitude and describe how our brain's goal-pursuing machinery works to give reinforcement power to an abstract thought.

Once a goal is selected by your nervous system, a kind of natural information cycle is set up to guide behavioral choices. There are four basic steps in this cycle. The nervous system must (1) hold the current goal in mind, (2) produce a critic signal for this goal, (3) use the critic signal to guide choices and improve the brain's model of the goal, and (4) select the next goal (or keep the current one active).

These four basic steps, when translated into their mathematical equivalents, form the basis of an approach to goal-seeking, learning, and decision-making called *reinforcement learning*. We touched on the idea of reinforcement learning in the last chapter, but here we are going to roll up our sleeves and dig into the details a bit more. Theoretical computer scientists and machine learning specialists are exploring models of machines that have goals, have internalized values, and use both to guide their decision-making. This theoretical work has had an important impact on our understanding of brain systems involved in drug addiction, goal-dependent learning, habit formation, and mental illness. It's a likely growth area between basic science and clinical science in the coming years.

———————

REINFORCEMENT LEARNING IS an approach to trial-and-error learning where a creature's actions are guided by a class of signals called rewards. These signals have been used in engineered systems and computer programs to equip these systems with goals, and to use the goals to guide learning. This is exactly analogous to the role played by reward signals in the brain. In engineering, reinforcement learning models are built into systems that program themselves through their interaction with the world, or at least a model of it. The main idea behind this approach is flexibility. Most practical real-world problems are so hard that a program without a lot of flexibility is doomed to be special-purpose. Ultimately, the goal of such systems is to produce autonomous, self-programming systems that achieve their goals flexibly and creatively.

The father of artificial intelligence, Marvin Minsky, wrote his 1954 Ph.D. thesis on reinforcement learning, and then promptly left the discipline. Until the 1980s, reinforcement learning systems were thought to be abysmally slow at learning anything useful, and this perception threw a dark cloud over the field. The problem with this assessment was that it was true! True, but not really relevant. No system can learn from scratch. No learning systems start without some assumptions about the problems they will face and how they might learn about them. A system with absolutely no assumptions could never learn. And biological systems are the best examples of "non-scratch" learning systems, since they all start with a deep biological lineage encoded in their DNA.

If a reinforcement learning system is trying to solve a problem starting with just a few assumptions, the learning is fiendishly slow and sometimes the system cannot solve even simple problems. Let's repeat the theme again: No sophisticated system, whether biological or man-made, starts from scratch every time it solves a new problem. This is especially true for biological organisms, which employ forms of reinforcement learning, but whose brains are built to solve specific classes of problems. A rat has little to no knowledge of how to hunt for shellfish underwater, but has an enormous behavioral repertoire for foraging in dark, spatially complex areas. So the rat comes pre-equipped with rich internal models of the problems that it will most likely have to solve to stay alive. Squirrels are terrific at burying thousands of nuts around a field and remembering the location of each. So the squirrel's spatial memory, like that of the rat, is absolutely terrific. But ask a squirrel to solve a simple puzzle or learn a few grammatical rules and it fails completely. The point is that it's important, even for flexible learning systems, to start with some general idea of the kinds of problems to solve. Once these assumptions are made, reinforcement learning systems can be extremely efficient, rapid learners. Backgammon provides a great real-world example.

Gerald Tesauro, a scientist at IBM's Thomas J. Watson Research Center, has used reinforcement learning to train a computer program called TD-Gammon to play backgammon. Versions of the program play against one another and use reinforcement learning algorithms to learn from their experience. TD-Gammon is now one of the best backgammon players in the world, and there is a direct analogy between the reinforcement learning used by TD-Gammon and that used by the human brain. Tesauro had several insights in training TD-Gammon to be a great backgammon player, but the most important one was realizing that the program had to play itself in order to get enough experience to be any good. Playing a human opponent took too much time for the trial-and-error style of feedback used in TD-Gammon.

TD-Gammon's exploits are fun to read about, but for our immediate purposes it embodies two key ideas—guidance signals and valuation functions. At the core of TD-Gammon is its valuation function, also called an evaluation function. Its central purpose is to provide an evaluation of the quality of each backgammon board layout and prescribe which move results (on average) in the likely best outcome. There are different parts to this function, but the basic idea is the same for each part: Use the reinforcement signal, the guidance signal, to learn the value of each board layout and possible moves from that layout. Our brains operate on similar principles—use experience and guidance signals to learn and store the value of different behaviors—a crucial requirement for being adaptive in a highly variable world. This kind of value learning equips us with a capacity to prioritize our actions and even our thoughts—some are more valuable than others and this information is stored in our brains.

Although the most recent incarnation of TD-Gammon is now world-ranked as a backgammon player, it differs from chess-playing programs like Deep Blue, which beat world chess champion Garry Kasparov in the late 1990s. Deep Blue emerged from years of development that employed a team of engineers, computer scientists, and a chess grandmaster. Among its many chess traits, Deep Blue could look at ~200,000,000 possible chess moves per second while Kasparov's brain was probably not capable of more than five or ten (if that)—and his brain is exceptional. Deep Blue was conceived, grown, and developed for one purpose only—to play chess. How could Kasparov even hope to keep up with such a remarkably speedy storehouse of chess-playing knowledge? He was not battling just a machine, but the integrated experience of a lot of other humans turbocharged to run millions of times faster than his brain. And yet Kasparov is still in Deep Blue's league. How is this possible? It's possible because of something that both Kasparov and Deep Blue share—evaluation functions.

Similar to TD-Gammon's, one core part of Deep Blue's program is its evaluation function. This is the part that decides how valuable a chess move is given a particular layout of the chess pieces. Now, chess is a much more complex game than backgammon, so building an evaluation function for chess is a vastly complicated problem. The evaluation function for Deep Blue is the "heart" of the program, and it's made of thousands of subparts, some focused on very specialized parts of chess games. But in a rudimentary way, the evaluation function acts much like the metaphor of one's "heart"—it determines what Deep Blue cares about. It cares about specific chess moves, but the analogy with an emotion should be kept in mind.

Kasparov also has valuation functions in his nervous system, but we're not that familiar with their implementation in his brain or what kinds of tricks they use. However, we can know that Kasparov's valuation function is crafty because it played chess toe-to-toe with a machine that could explicitly look at solutions 100 million times faster than he could. Think about this for minute. It's like having an impossible footrace with a rocket and miraculously the race ends up nose-to-nose. While we're not sure of the detailed tricks used by Kasparov's nervous system, we know three things that differed tremendously from Deep Blue—efficient representation, valuation, and learning.

Kasparov's brain already possessed efficient ways to represent and value the game of chess, and these representations could adapt as play went along. And all this was done with extreme efficiency. How do we know? Because Kasparov did not burst into flames; he remained only warm to the touch. And as wacky as that might sound, Deep Blue's computers could easily have burst into flames (or at least melt down) and their processors were never just warm to the touch. The computers running Deep Blue's program were terribly hot, requiring fans and heat-dissipating equipment to carry away loads of wasted heat. So just on grounds of efficiency, Kasparov's valuation functions were vastly better than Deep Blue's.

Kasparov's valuation functions are better than Deep Blue's for another reason: They can be flexibly applied to other problems. Kasparov can make analogies between chess situations and problems that he faces in other areas of life, and use the connection to reason in new ways. Deep Blue is chained by its program to one destiny—play chess and only play chess. It has no capacity to even sense a bigger world outside of chess games. But of course Kasparov does. So while we should all admire Deep Blue for the tremendous accomplishment that it represents, biology still has much to teach us about flexible and efficient intelligence. And our ignorance in this arena is expressed further by our inability to describe what exactly makes Kasparov's brain so good. The description of Kasparov's play against Deep Blue relies on an impoverished vocabulary, and he is typically described as "intuitive," "competitive," and so on. If we knew a little more about exactly how his brain evaluated board positions, we might have a richer way to describe his performance. And our descriptions would almost certainly focus on his amazing valuation functions and the deep tricks they embody. Although outgunned by the raw computational power of Deep Blue, Kasparov's brain found a way through the same problem and it was efficient.

Returning to Gerald Tesauro's efforts with backgammon, his

program was trained initially without any expert but only with a critic signal like the one described above, a signal that simply guided play rather than prescribing what it should be in each and every circumstance—like the initial brute-force method used by Deep Blue. Guidance beats prescription. Guidance is great when flexibility is required. Prescription is great when the solution to a problem is known once and for all. Creatures need guidance, not prescription. Why? Because the most constant feature of a mobile creature's environment is its inconstancy, its raw variability. Rocks, roots, holes, food, enemies, etc.—they are all extremely variable, and guidance is a better strategy than trying to prescribe every possible solution for every possible contingency. Life is complicated and tough, and real creatures need learning mechanisms that know how to stumble, get up again, and learn from the experience.

Reinforcement learning is a hot area of research, with practical and theoretical results expanding in number each year. However, for insights into our own brain's reinforcement learning systems, leading-edge theoretical results are not needed. Instead, "starter-kit" reinforcement learning algorithms now shed light on how brains generate goal-directed feedback signals. The importation and application of reinforcement learning to psychological and biological problems was cofathered in the early 1980s by the work of Richard Sutton and Andrew Barto. These two scientists hail from different scientific cultures, Sutton from psychology and Barto from computer science, but they were drawn together by the idea of systems that learn from experience. A similar, but parallel, influence flows from the work of Bertsekas and Tsitsiklis of MIT from the engineering side. Both duos shone light on these problems after a doldrums period in the 1970s when it was common wisdom that reinforcement learning systems were painfully slow. In both camps, the basic step forward was to realize that if you know a little bit about a problem beforehand, then a reinforcement learning system does not have to start out maximally ignorant, no "blank slate," and in these cases reinforcement learning systems can be made quick and efficient.

All animals start out life this way; they all begin with rich models of what they must learn and roughly how to do it. This doesn't mean that the content of their learning ends up the same. A ferret growing up in the wild ends up solving different problems from one raised in my living room. They both continue to learn the way that ferrets do; however, each ferret learns to solve different problems. Each environment puts its stamp on the content of the ferret's experience, but raising a ferret in a living room will not magically change its learning strategies. Now, breeding ferrets, while selecting for certain problem-solving abilities, may indeed change learning strategies, but here we're focusing on how animals use reinforcement learning moment to moment to guide behavior. Without a lot of hard-won assumptions about the learning problems, all learning systems, including biological ones, would be rather helpless. We won't analyze this position in any depth, because it would take us too far from the main points of this book. These ideas have been analyzed and discussed elsewhere in excellent books. But in biology, there are no true blank slates. The idea that a mobile, adaptive creature starts as a blank slate is contradicted by almost as much evidence as there is to support evolution by natural selection. This holds for rats, dolphins, worms, humans, bacteria, and so on. Let's turn our attention to the information cycle that uses goal selection and critic signals to guide our decision-making and learning.

For humans, goals can be as immediate as food or sex, or as

abstract as the feeling of having a new career, winning some yet-to-be-held contest, or warranting praise from colleagues for solving a difficult problem. Despite their differences, all goals have one thing in common: They can all be used by our brains to direct decisions that lead to the satisfaction of the goal. We would all love to find satisfaction as the endpoint of most of our decisions, but ironically we may be designed to keep true satisfaction just out of reach. Psychiatrist and neuroscientist Gregory Berns has taken to task the issue of satisfaction and its relation to the brain's reward and guidance systems. His broad conclusion is that total satisfaction is designed to be just beyond reach, and for a good reason. It makes for a better learning machine. We have no incentive to stop learning. His book on the subject makes fascinating reading.

Modern brain research has identified a kind of trick used by the brain to define goals, even abstract ones. The simplest goal can be pictured as the physical-state brain of your brain in response to some rewarding event like eating food, drinking water, or having sex. Our brains are designed to seek goals by following the advice of critic signals (guidance signals) that report the deviation of our actual experience from our internally represented goals. The basic idea is that our internal goals get compared to real or imagined (simulated) experience to produce ongoing criticism in the form of identifiable critic signals. There are separate brain mechanisms that use this criticism to adjust behavior or to learn (store information). And these guidance signals never rest—they guide learning continuously and unconsciously. Automatic, guided learning can take place online, while experiencing the external world, or offline, while simulating the world (dreams or imagination). Models from reinforcement learning helped identify the true nature of these biological critic signals, and these same models are now directing experiments in humans that seek to understand the scope of the critic's influence.

Kind of creepy—critic signals whispering to parts of our brain, and all underneath our conscious awareness. It's important that the words *critic* and *guide* are used to describe these neural signals. Our brain treats them the way anyone treats a critic—it ignores a lot of the advice. Some suggestions are more informed and useful than others. This advice, the critic signals, consists of two basic parts: a piece representing immediate experience (feedback) and a piece carrying long-term judgments about the future. These critics (and there are many) direct our choices in a kind of abstract warmer-colder game, just the way our Alaskan guide example from before gave us little local commands about how to "stay safe in the wilderness."

In the warmer-colder game some object, let's say a dollar bill, is hidden in a room by one child (the guide), and another child (decision-maker) is directed to this object by following the warmer-colder criticisms issued by the guide. If the child moves closer to the hiding place, the guide sounds off, "Warmer," and if the child moves away from the goal, the guide says, "Colder." Pretty simple, but fun, especially if you can store a mental model of where you have been and can remember the series of warmer-colder critiques as you moved from one spot to the next. If you're really good at this game, you will form a kind of temperature map of the room that emanates from your starting place. This map, built by experience, can be used to find the hidden object. It would consist of a bunch of little colored arrows, each arrow indicating the direction of movement and the color indicating whether the movement caused the temperature to go up or down. The color scheme for the warmer-colder critiques encodes the *value* of the actions available at each position in the room. And each action is the direction of the arrow. Altogether, this collection of information forms a simple *valuation map* of the room. It gives the relative value (for reaching the dollar bill) for each move at each position in the room.

In the real world, things are not this neatly presented, the problems are often more abstract, and simple warmer-colder critic signals would be woefully insufficient. But the brain has a few more tricks up its sleeve. The brain's critic signals are more complex, there is more than one critic ranking different aspects of each action and even keeping multiple competing goals active. This ability to compare goals, rank goals, and keep multiple goals in mind depends in large part on our expanded prefrontal cortex. We will focus on critic signals, but the functions of the prefrontal cortex are an important subject because it connects some of the most interesting aspects of our mental lives to neural function.

And the brain also has a crafty redeployment trick available to it, one that is directly related to the superpower mentioned earlier. It can use critic signals designed to find food and water to pursue abstract ideas like social goals. It does this by reusing an evolutionarily old reward-chasing system, but plugging some abstract idea into the "reward" slot. It's this last feature that allows ideas to act as rewards.

In the brain, the critic signals act much like the warmer-colder critiques just described, delivering information about whether things are getting better (warmer) or worse; however, they contain more information than mere warmer-colder assessments. The brain's critic signals do not possess a perfect "bird's-eye knowledge" of a goal; instead, they use stored experience (memory) and rich models of similar problems to make educated guesses about the *value of possible future actions*. As we said, these critics combine two important sources of information: information about immediate reward (feedback from "what I'm experiencing now") and judgments about future reward ("what I'm likely to get in the long-term future") to produce a kind of *smart error signal*—driven by the present, informed by the past, and guided by the likely future.

The critic systems broadcast these error signals to multiple regions of the brain using long neural connections (lots of wire), where they can be used for different purposes. This is an example of another efficiency built into our brains. These critic signals, encoded as electrical impulses, travel along some of the longest axons in the brain to deliver their error signals; consequently, they consume a lot of bandwidth per message sent. We should not be surprised that they carry information that is used in at least three important functions: goal selection, learning, and decision-making. They also carry other kinds of information—just as the DSL modem or cable modem in your house piggybacks signals from your computer on phone and cable television circuitry. Someone noticed unused bandwidth and decided that cable and telephone companies could use it to make money by sending packets of information back and forth to our home computers—an increase in efficiency. Just like the modems, the brain takes advantage of unused bandwidth by packing multiple signals onto the same wires.

We can also describe these critic signals in the language of reinforcement learning. The goal of the learner in reinforcement learning is to use past experience and local feedback to maximize the total future reward acquired in the future. It's just like a command one might expect from a biological banker, to "make choices that will maximize returns over your lifetime." Depending on the environment and the behavioral repertoire available to the learner, there are many ways to solve this problem—another reason why advice is needed and not prescriptions. For most mobile organisms, reinforcement learning systems have critic signals tied to the acquisition of food, water, sex, and social exchange—commodities critical to survival.

Let's break this description down into its components and sketch a picture of how these systems compute their advice. All reinforcement learning systems have three major parts: (1) an immediate reinforcement signal that assigns a number to each state of the creature, (2) a stored value function that represents a judgment about the long-term value of each state, and (3) a policy that maps the agent's states to its actions. The immediate reinforcement signal is the myopic part. It reports how good or bad things are right now. The value function stores the long-term judgments of the current state based on past experience: "The last time I was in this state, the future that followed was really quite good and yielded a lot of reward." But instead of using English text to carry their message, these systems encode "last time," "really quite good," and "yielded a lot of reward" into numbers represented in your brain as patterns of neural impulse activity. The critic signal combines (1) immediate reinforcement information with (2) changes in value to produce what is sometimes called a reward-prediction error signal. The reward prediction error could represent errors along different behavioral dimensions, for example, errors in the amount of food experienced or even errors in the social feedback expected during an exchange with another human. At one moment, the critic may be generating errors about food and the next moment it may be generating errors about expected social responses. It's truly an important multipurpose system, so there is no single answer to the question "What information do these critic signals represent?"

And what's more important is that real experience and "simulated experience" can both generate critics. The critic signals can be generated through actual exploration of the world or through "mental exploration" of events that have not happened. So even contemplated actions can engage these systems, and it's this capacity to direct mental exploration that allows the critics to guide innovative thinking. Now let's get concrete. How can these abstract descriptions of guides and critics and reinforcement learning be related to neurons, synapses, and neurotransmitter systems?

These descriptions of reinforcement learning systems actually match the behavior of specific neural systems in the mammalian brain. One of the most important such systems is the dopamine system and the role that it plays in learning about rewards and directing our choices that lead us to rewards. This system is used for simple things like associating a visual stimulus like a Ben and Jerry's sign with the rewarding experience of eating cold, sugar-laden ice cream (possibly peppered with nuts and crushed candy bars and . . .). These same dopamine systems can be redeployed to guide decision toward more abstract rewards like money or even pleasant social feedback.

DOPAMINE NEURONS FORM PART of a collection of neural systems in the brain called *neuromodulatory systems*. The term derives from two important properties that these systems all possess: (1) They act at fairly slow timescales compared to other processing in the brain, and (2) their axons project to widespread target sites throughout the brain and spinal cord (they broadcast their computations). Together, these easy-to-spot features conspired to convince scientists early on that neuromodulators provided sluggish *volume control* or modulation over widespread regions of the nervous system, like a slow broadcast system. This view is correct, but we now know that it's woefully incomplete. Neuromodulatory systems also compute much faster timescales and they carry more than just simple volume commands—"turn things up" or "turn things down." Instead, we now know that there are lots of signals encoded in the activity of these systems, and some of it has been decoded.

Our efficient computational principles from Chapter 2 should advertise these systems as important and efficient: slow as they can be and important because so much wire has been spent on them. The majority of their connections originate at the ends of long-range axons, the most expensive communication channel in the brain. Neuromodulatory systems indeed play a volume-control role, but they also direct ongoing valuation of experiences, decision-making, and even changes in working memory. Dopamine is not alone in these roles. Other neuromodulatory systems share these properties, but differ in the neurotransmitter chemicals they employ to deliver their messages to target regions of the brain. These include serotonin, acetylcholine, norepinephrine, histamine, and others. All these neurotransmitter systems are the subject of intensive research, but here we focus on dopamine, since it provides a concrete way to understand how ideas can acquire superpower status.

Why should we be interested in the dopamine system? The dopamine system is hijacked by every drug of abuse, destroyed by Parkinson's disease, and perturbed by various forms of mental illness. Overstimulating the dopamine system causes dramatic cognitive delusions, including auditory hallucinations and the erosion of our sense of self. These are just a few reasons that so much hard work is directed at understanding dopamine systems. Most importantly, dopamine systems implement a form of reinforcement learning that is now beginning to yield some of its secrets and may serve as one example of how the human brain integrates emotional states with rational thought, a key feature of being human.

Dopamine neurons are clustered in little islandlike arrangements in the brain stem (midbrain), a structure that we share with all vertebrates. These islands are distributed in a roughly symmetric fashion across the midline of the brain stem, just like the symmetry between the two sides of your face. On each side lives a collection of 15,000–25,000 dopamine neurons that send out long axons to widespread regions of the brain. This is a very small number of neurons when one considers that there are approximately 250,000 neurons under every square millimeter of the cerebral cortex and a full 100 billion neurons throughout the entire cortex. The midbrain dopamine system is a broadcast system where important goal and guidance information is sent to multiple brain areas at once (the broadcast). These small islands of dopamine neurons provide the only source of dopamine to the cerebral cortex; if you lose or damage them through drug use, injury, or disease, that's it, no more dopamine to the thinking parts of your brain and the parts that select actions.

DOPAMINE NEURONS COMPUTE and distribute one of the brain's most important critic signals, and they represent the best biological example of a set of neurons capable of computing a reward-prediction error signal. These neurons encode this reward prediction error as bursts and pauses in their production of electrical impulses, which are then distributed throughout the brain by their extensive axons. This encoding is easy to picture. Bursts of impulse activity mean "Reward is more than expected," pauses mean "Reward is less than expected," and no change means "Reward is just as expected." These neurons are emitting information all the time, even when they don't change their baseline impulse rate. For fifty years, the baseline impulse rate for neurons was thought to be like idling in a car. The engine is running and exhaust is being produced, but the car is not going anywhere. But dopamine neurons appear to be broadcasting information all the time—even when they are idling. This is a big change in the way neuroscientists think about idling behavior in neurons, but we would have expected this from our principles of efficient computation in Chapter 2.

Our modern understanding of the fast dopamine signal rides on the back of twenty years of remarkable experimental work by the neurophysiologist Wolfram Schultz. His discoveries have played a central role in our current theoretical understanding of this important system. In fact, without his work, no theoretical description would have been possible. Like many others, Schultz records the electrical impulses emitted by dopamine neurons in the brains of monkeys while they are carrying out reward learning tasks. Here, *reward* means something like a squirt of fruit juice or a piece of apple. The recordings depend on placing a microwire right next to the dopamine neurons in the same way recordings are made in human brains to monitor for epilepsy or plan a neurosurgery. There are no pain receptors in the brain, so the recording part itself is painless.

A lot of early work on dopamine neuron activity focused on relating it to movement. Why? Because of Parkinson's disease—it kills dopamine neurons while making you stiff and inhibiting voluntary movements. So dopamine neuron activity should encode some aspects of movements, right? Wrong. During the 1980s, lots of laboratories around the world were interested in exactly this question, but none found a systematic relationship between dopamine neuron activity and movements, Schultz's lab included. However, as in the case of all big discoveries, Schultz began to notice something, an event that heralded more than a decade of new discoveries from his laboratory about dopamine neurons. He had a moment typical for science, not a big-bang "eureka" moment, but something odd caught his attention. As the writer Isaac Asimov has so astutely pointed out, "The most exciting phrase to hear in science, the one that heralds the most discoveries, is not Eureka! (I found it!) but 'That's funny . . .'"

Schultz noticed that dopamine neurons changed their activity when "important" events happened, like a juice squirt, or the appearance of food, or even a sound in the laboratory that predicted that food or drink was about to be delivered. He got suspicious. Maybe dopamine neurons encode information about rewarding events and the events that lead to reward? In fact they do, but Schultz's work already suggested that it's not as simple as "dopamine = reward" or "dopamine = pleasure." The electrical responses of dopamine neurons are confusingly complex, especially when rewards are present. Adding to this difficulty, the dopamine neurons learn from experience; they adjust their electrical impulse rate depending on the history of the organism's reward experience. For several years, Schultz characterized these odd and flexible reward responses in dopamine neurons, each experiment cleverly testing some new aspect of the dopamine responses. But in 1991, the clever experiments of Schultz met the clever mind of Peter Dayan, an English mathematician who had worked previously on theories of reinforcement learning. Dayan also had an "Asimov moment"—his "That's funny" experience came when he recognized a striking resemblance between dopamine neuron activity and error signals used in abstract reinforcement learning algorithms.

Dayan, at the time a postdoctoral fellow in the Computational Neuro-biology Lab at the Salk Institute, worked with me and Terry Sejnowski to develop a simple reinforcement learning model to account for Schultz's remarkable data. It was an amazing match. The model showed that Schultz had discovered one of the central critic systems in the mammalian brain, and one that encoded its criticism in the delivery of dopamine.

The recognition that dopamine delivery acts like a critic signal was a breakthrough. Several issues cleared up immediately. There was now no simple answer to "What does dopamine encode?" Even though we know it can represent reward prediction errors, these could be errors along any dimension that a creature's brain frames as rewarding, even something as abstract as the idea of "a spaceship waiting to take you to paradise." At one moment they may represent the difference between money expected and money acquired during a gambling game. At another moment, they may deliver a very different error—like the difference between the expected experience of some food and the actual experience. We can no longer use the blunt idea of dopamine equals reward. Fast dopamine fluctuations in our brain do not represent just one thing, nor do they play just one role. Instead, we see one physical signal carrying multiple kinds of information. Now it's clear why a complete understanding of these relationships requires a marriage between computer science, cognitive psychology, and neurobiology. As we will discuss later, it also needs other parts of social science, like economic theory and social psychology. What else should we expect? This system influences important mental features that make us human, and we're pretty complicated.

And this reward-prediction error model connected dopamine function to valuation, a connection that offers a new way to think about Parkinson's disease, drug addiction, and various thought disorders. We address these in the next chapter, but from this new point of view. So dopamine is doing lots of error jobs, depending on how one person's brain frames a situation—context matters to this signal. The model also anticipates the outcomes of experiments not yet carried out, and indeed it has been used to design all kinds of new experiments in human subjects. It is being applied to populations of subjects afflicted with the same maladies that make dopamine such an important neuromodulatory system to understand. In recent years, the dopamine critic signal has been subjected to extensive quantitative analysis by Hanna Bayer and Paul Glimcher at New York University. This work has corroborated Schultz's basic finding about the dopamine reward prediction error. Altogether, it's now pretty clear that rapid changes in dopamine signaling in the brain encode a signal almost identical to that used in artificial reinforcement learning systems, thus opening the door to a variety of new theoretical analyses of brain function. Remember, this is the same critic signal used to train TD-Gammon, now the third-ranked backgammon player in the world.

So how does the dopamine system contribute to the superpower? The answer is related to a remarkable trick that the mammalian brain long ago discovered.

Ideas act as reward signals from the point of view of the prediction error systems.

Ideas play the role of reward signals fed to the dopamine neurons, which use information from other brain regions to predict this new "idea reward." In the process, the ongoing output of the dopamine neurons becomes an error signal that tells the rest of the brain how to adjust its decisions and learning to acquire this "idea reward." Rapid changes in dopamine delivery guide learning and decision-making so as to acquire this "idea reward." As long as the idea can maintain its status as a "reward," the rest of the brain adjusts itself to learn about this reward just as it would adjust itself to learn about a new source of food, water, or sex. Now, of course, this is an extremely simplified version with gaps, but that's the basic picture.

Ideas gain the power of rewards and become instantly meaningful to the rest of the brain, especially the learning and decision-making algorithms running there. Now, this kind of trick provides for an extremely creative learning machine. It can choose to ignore its instincts momentarily and pursue a thought to the exclusion of everything else. It is easy to see how such a power could be useful for generating cognitive innovations. An idea with the beckoning power of ice cream can control a succession of thoughts for some time. The effect is just like foraging for food hidden under rocks and behind bushes in a field. The food acts as the goal, the dopamine system computes an ongoing food prediction error, and the ensuing dopamine fluctuations guide the actions selected to obtain the goal. Same effect with an idea, but this requires a way to forage through a field of ideas, an "idea space," the same way one would forage for food. Cool trick. Redeploy foraging in the pursuit of cognitive innovation. Foraging fields for food becomes foraging a mental storehouse for new ideas.

However, this innovation process must be tightly controlled. Everyone can't simply sit around humming away while they overdose on their latest and greatest idea as though they were sucking on a succulent fruit or eating a rich meal. This kind of behavior would be grossly inconsistent with survival. There must be lots of checks and balances in such a process. Two important limitations come to mind right away. First, there must be some kind of filtering process that circumscribes the kinds of thoughts that can act as a reward signal. Second, the control must be self-limiting, that is, it must be relatively short-lived. Together, these two limitations predict the existence of a mechanism that limits "active time on the reward system." Although an idea can gain the status of reward for a short time, the system should adjust quickly to cause the idea to lose its status as a reward. This mechanism would keep pathological ideas from literally running away with your body—which is what happened to the Heaven's Gate cult. The idea of self-limiting reward status is currently just speculation, but there is evidence in the dopamine system that reward status is very adaptable.

Just consider someone trying to solve some difficult mathematics problem, for example, the much-publicized work of Andrew Wiles, who solved Fermat's last theorem, formerly one of the "Holy Grail problems" in mathematics. Many famous mathematicians tried and failed to solve this problem, but Andrew Wiles succeeded. In the published accounts of his efforts, he is reported to have worked on the problem for about seven years. What are we to make of this? The solution to the problem was extremely important to him, especially so since the endeavor was professionally very risky. There was a substantial likelihood of failure, and although lots of good mathematical results might emerge during the chase for the solution, I'm sure that the pursuit seemed to Wiles a risky one. Our point here is that "solving Fermat's last theorem," as an idea, had to gain value in his mind in order for him to reorganize his life and his moment-to-moment behavior in order to try and acquire this goal. "Solving Fermat's last theorem" was literally a reward to him, just as it was to every mathematician who tried and failed to solve the problem. And I suspect that there must have been dark moments in this pursuit when all seemed lost or at least immeasurably murky. I don't know Andrew Wiles and this may not even be close to his conscious experience of his work on this problem, but the simple fact of his having worked on it for seven years is a sufficient criterion for me to call it rewarding.

WHERE EXACTLY DOES VALUATION FIT into my descriptions of neural reinforcement learning, the potential superpower status of ideas, and the way that a critic signal directs learning and decision-making? Surprisingly, this is best addressed with a simple example from the experiments of Wolfram Schultz. The reward-prediction error signal (critic) was first identified using experiments in monkeys, where lights predicted the future arrival of juice squirts a short time later. When the light is initially presented, there is no change in dopamine neuron activity; that is, the light is initially associated with a "things are just as expected" signal in the neurons. However, the arrival of the juice a short time later causes a burst of activity in the dopamine neurons; that is, the juice delivery is initially associated with a "things are better than expected" signal in the neurons. If the light-juice pairs are repeatedly delivered to the subject, two remarkable changes occur. The initial response associated with the juice, the "things are better than expected" response, goes away. Poof! It literally disappears and the neurons no longer change their activity when the juice arrives following the light. This means that the dopamine system has learned to expect the time and amount of juice delivered, and it demonstrates this by not changing its activity ("things are just as expected") when the juice actually arrives. Now for the second change. The initial nonresponse to the light, the "things are just as expected" response, changes into a burst of activity. The neurons now report that the light is "better than expected"; that is, they react to the light in the same fashion that they reacted initially to the juice. In computational terms, the value of the juice has been transferred to the light, which becomes a "value proxy" for the future juice delivery that it predicts. The neural response to the light becomes a kind of neural promissory note for the real value (sugar water) that will soon arrive.

This shift of the burst response from the juice to the light is communicated by the widespread axons of the dopamine neurons telling the rest of the brain to treat the light as though it carried the same intrinsic reward value as the juice—basically a proxy assignment scheme. The light's consistent relationship to the juice gives your brain permission to let the light act as the proxy for juice, a value proxy, just like an option on a stock—something that stands for something that has value in the future. So the brain is solving a statistics problem here—it's constantly looking for stimuli that predict future value and by doing so the brain can use these "value proxies" to make better decisions about the future.

The alternative is to use a reactive system that must wait until valuable events happen and that then generates a reactive decision to them—such a wait would often prove fatal. The value proxy scheme actually employed by the brain makes it an anticipatory system rather than a reactive one. And here we see a connection back to the efficiency arguments made in Chapter 1: Anticipation allows the brain to build and employ models of the future value of actions that it is contemplating. It has a kind of *value map* of future possible actions, an invaluable device for a survival machine that compares the relative value of future potential actions. "Do I chase that potential mate or wait around for something better?" "Should I run from the predator that I think I hear in the distance or should I continue eating away at my latest kill?" A more cognitive version of these behavioral conundrums might read, "Am I more afraid of the risk associated with putting my pension in the stock market or my possible future pain if I don't and the market goes up?" Your valuation systems carry out such comparisons by providing a common neural currency in the form of value proxy—and of course this is just what money is: something that stands for future value (when you then trade it for something else).

In the simple light-juice pairings above, the transfer of value to the predictive light is just one symptom of how sophisticated the value proxy system is in the human brain. If there was a bell that consistently preceded the light, then the burst of activity would transfer all the way back to the bell, passing the light proxy's value back to the bell. After this further transfer, the dopamine system will only give a burst ("better than expected") to the bell and will not change its activity to the subsequent light or the juice. This, too, is efficient. Since both bell and light predict the same future reward, why bother making extra, energy-consuming impulses to both predictors? This would be redundant. OK, so why pick the earliest one? Efficiency again. This gives the system more time to prepare its actions, hence more time to avoid costly, energy-wasting decisions.

Now let's generalize a little more. Any realistic learning scenario is extremely complex, with lots of possible stimuli predicting future (valuable) outcomes. However, if the brain starts out with a good model of the learning problem, that is, what it should generally do in a particular situation requiring learning, this form of reinforcement learning is quite rapid and efficient. It permits a self-programming creature to transfer value to events that predict future valuable events. This transfer endows the system with the capacity to anticipate the value of possible futures, and to make good decisions now that will lead to brighter, reward-rich futures later. In later chapters, we will explore imaging experiments with humans where their internal capacity to convert stimuli to a common currency shows up in a collection of common brain regions and responses.

Let's point out an important and perhaps obvious chronological fact—the mammalian brain possessed value proxies well before humans existed and, consequently, well before humans discovered the idea of a currency. And this capacity was not new to mammalian brains, but existed in varying degrees in all mobile creatures that must assign value to their potential actions and choose those that are likely to yield the best returns. There is a deep efficiency principle at work with the value proxy scheme in our brains—the dramatic reduction in transaction costs achieved by using a common medium of exchange.

Consider the following thought experiment. There is no medium of exchange (no money). You grow wheat and someone else a hundred miles away harvests shellfish from the sea. How costly is it for a trade of wheat and shellfish to occur? We will answer this question lightly. The two of you gather up bushels of wheat and baskets of shellfish respectively and travel fifty miles each to a prearranged trading site. At the site, possible trades are "simulated." You offer two bushels of wheat for each basket of shellfish and your trading partner wants three. The deal settles on two and a half, the trade is made, and you lug your remaining bushels back home; things didn't work out evenly. The need to meet at a common site and carry the goods with you represents enormous direct and forgone costs. Once your caravan has embarked, you are forgoing other deals that you might have otherwise entertained. The notion of a common currency lubricates this entire scheme, especially in terms of lowering simulation costs—the cost of simulating possible trades and assigning values to them. If you and your potential trading partner can agree on some common currency that's low-cost (light and easy to carry), then you can simulate possible trades, establish its value to you and your partner (remember two are participating here), and convert it to the common currency. This allows many different trades to be simulated beforehand and the relative value of each, as expressed in the common currency, to be known. Basically, you can go to the trade well prepared, and once there you can easily compute the value of any new arrangements that you have not already valued. A common currency scheme only works if there are checks and balances built into the brains of the traders. The most important check is trust, the ability to rely on one's partner to obey the rules, whatever they may be.

Neural currencies permit this same general scenario, except more quickly and covertly. The construction and use of neural currencies represent a cost-saving maneuver on the part of the nervous system, harkening back to its need for extreme efficiency. In Chapter 1, we saw that the main business of the brain was modeling—modeling itself, modeling its sensory experience, remodeling its past, and modeling its likely futures. But the brain must be able to decide which models are more valuable to it. It must possess a dynamic system for placing models and their predictions on some common valuation scale in order to choose the model deemed most valuable for the moment. The need for ongoing valuation applies to models of predators, prey, and, in the case of humans, other people.

How should the brain value other humans? This is really a loaded question. At one level, it's an economic question now underwritten by a growing understanding of how our brain assigns value, the dynamics of this process, and the parts of the brain involved. Alternatively, this is a chapter about the superpower, that is, the capacity of ideas to accrue the value of primary rewards. This means that we can and often do rise above (or sink down to) the level of our instinctual social responses. Our new and growing understanding of the brain's role in the process of endowing ideas with "reward power" should provide a window of hope into our understanding and misunderstandings of human behavior. This knowledge is itself an idea that can accrue value in our minds, perhaps shaking us out of some of our prejudices and otherwise maladaptive social practices. Of course, this does not mean that we will ever erase our impulses, only that our understanding of our brains can help us recategorize our impulses and provide a way around some of our worst instincts.

THE VALUE MACHINE

And the Idea Overdose

It is impossible to get out of a problem by using the
same kind of thinking that it took to get into it.
ALBERT EINSTEIN

ONCE LIFE STARTED to move, valuation mechanisms were an inevitable consequence. A creature that moves left one moment forgoes what it might have obtained had it moved right. This is an economic decision and it selects creatures that make more informed choices. We're back to the Energizer Bunny problem—recharge. Even simple, single-celled creatures must act like rapid-fire economists. They collect their own energy, store it, and use it later, and all these processes take energy. Collecting stuff over here means not collecting stuff over there. Being distracted by a "bad" food source could mean a superior food source is missed. The major source of values in these tiny economists is movable bodies. Creatures experience the external world at specific space and time scales, a fact forced on them by their bodies, which have particular sizes, speeds, and styles of movement. This is why the wisdom of good economic valuation was forced upon all the mechanisms that make life run.

Take the specific example of *Escherichia coli,* a rod-shaped bacterium about one to two millionths of a meter long and about one half millionth of a meter in radius. A little guy. These creatures are so small that for them, moving through water is about like a human trying to swim through molasses. You really need to choose your motion carefully, since it takes some serious energy. If *E. coli* were much smaller, they wouldn't even need to move to acquire food because it wouldn't be worth the effort. One of their favorite foods is the amino acid L-aspartate (not to be confused with the artificial sweetener in the blue packets, which also contains aspartate).

Even single-celled *E. coli* are capitalists; they follow the money. They follow the aspartate, a source of energy and raw materials. *E. coli* can sense changes in aspartate concentration, and using a smart propulsion system, they move toward food and away from poisons, a process called chemotaxis. The propulsion is generated by rotating hairlike extensions called flagella (about six or so per cell) that are several times their body length. When rotated counterclockwise, the flagella come together to form a single propulsive "ponytail," and the bacterium runs along forward. When the rotation reverses to clockwise, the ponytail flies apart, forward motion ceases, and the bacterium tumbles and reorients in a random direction. That's it—just run and tumble. When they bang into less aspartate, they tumble more frequently, and when they bang into more aspartate, they decrease their rate of tumbling and run forward more. Together, these two simple behaviors allow *E. coli* to do a kind of wobbly "drunk's walk" up concentration gradients of aspartate. The decision they make is to choose the rate of tumbling, and it's the aspartate encounters that determine that rate. *E. coli* builds a kind of microbe's picture of its "external aspartate world" using its run and tumble strategies. It has no neurons; remember, it's just one cell with just a little bit of DNA in it, but it's a remarkable piece of engineering. Without neurons, *E. coli* builds its aspartate worldview using internal biochemical signaling networks. In fact, from the moment of its "birth," *E. coli* comes preequipped with the capacity to build such world models.

The details of the connection between aspartate sensing and changes in the decision to tumble make a beautiful scientific story, but our purpose here is to emphasize the *valuation function* built into this scheme. It should suffice to say that *E. coli* has committed a large number of its internal resources to build, value, and respond to a model of the "aspartate world" around it. Why is it doing this and how does it benefit from this kind of investment? The answer lies in considering what would happen to *E. coli* if it could not build such internal pictures of the external world.

A hypothetical "dumber" version of *E. coli*, let's call it *D. coli*, might not have spent its energy and time this way, and instead just consumed the aspartate as it was encountered. What's the problem with simply eating what you find? The problem really has to do with how hard it will be to find food "tomorrow" and the need to live a little while longer—at least until you reproduce. If aspartate were always available in excess, *D. coli* might well be more adaptive than *E. coli*, since it doesn't waste energy trying to build "aspartate models" and control its behavior to run toward gradients of this energy source. However, *D. coli* is "dumb" because the real world is simply not that accommodating. External energy sources aren't uniformly distributed, nor are they always plentiful. *E. coli*'s energetic investment into model-building is sensible, and amounts to splitting the net energy from each aspartate molecule consumed into two separate streams: one for *fuel* and one for *information* to be used to build a better model of its "aspartate world"—a *modeling stream*.

The modeling stream is a protoversion of a reward signal, since it has value to the organism beyond fuel content. The bacterium has a value function built into the way it handles its aspartate encounters, and based on this valuation, it makes decisions about running and tumbling. My point here is conceptual and not mechanistic. I'm emphasizing that the *E. coli* had an early economic incentive to judge accurately the value of its aspartate encounters, and to use some fraction of them as an information source rather than just a fuel source. This takes energy, and so a bet must be placed by the investing organism. Here, the bet placed by *E. coli* is that the cost of the information stream will return to it in the form of better aspartate yields in the near future; yields that are greater than or equal to the investment cost for building its models and directing its behavior. This investment becomes a measure of the degree to which *E. coli* cares about the aspartate environment around it. It's not too hard to imagine how such a fuel and information processing division could have evolved early on. Now, each bacterium's decisions aren't really as animated, first person, and anthropomorphic as my description suggests, but neither are all of ours. No matter how sophisticated our own decision-making capabilities, they must continue to solve many of the same basic survival problems encountered by *E. coli*.

The analysis here reveals that valuation is a basic need of all mobile organisms. Unlike the burning of fuel to drive motion or digestion, it is instead a process that consumes fuel now in order to build better models of an uncertain future. Valuation is a cost-assignment mechanism, a prerequisite for an algorithm or machine to care about anything. Our detour into the decisions made by motile bacteria should emphasize how fundamentally value mechanisms are connected to life itself. As I pointed out in Chapter 1, an efficient computational device must have goals, it must care about something; otherwise it has no standards to guide its behavior. In a way, we've seen that credo expressed here. Simple forms of life must care about something, even if their "caring" is rudimentary by current human standards. At this level, caring can be analyzed as the economics of committing to an information-bearing signal within the organism, a signal that the organism must pay for as an "up-front" energetic investment. At the level of humans, we saw in the last chapter that the superpower can redeploy valuation mechanisms to give us social agency and to give some ideas the same behavioral power as primary reinforcers like food and water.

I should emphasize that the signals that I'm describing here and throughout most of this book should be conceived of as computational signals, *not subjective feelings*. I suspect that an enormous amount of our mental machinery can be captured in a value-based computational theory of mind, but not all of it. I'm not so sure about feelings, and indeed much has been written on this subject. However, I should also point out that social agency, the ability to distinguish oneself as an individual separate from the rest of the world, is a feature that does not require feelings to operate and this possibility is an open area for study. With the evolutionary origins of value mechanisms sketched out, let's return to human goal-setting and goal-sustaining mechanisms.

The bacterial example shows that one can make headway on problems of the origins of value by looking squarely at some of the simpler answers (creatures) to evolution's questions. For humans, valuation is often a hidden function. Unlike bacteria, humans can pursue complex goals that require extended sequences of different actions, sequences that can extend over great periods of time. Therefore, humans must possess the ability to "keep a goal in mind" within the nervous system while it is being actively pursued. Research repeatedly points to the prefrontal cortex as a site where goals are formed, selected, and actively maintained. This work is far too broad to review here, but remember that compared to other species humans possess a dramatically expanded prefrontal cortex. My purpose is focused—to examine how a goal gains rewardlike status and guides choices—while restricting the discussion to those features that can be captured in computational models and turned into equations or simulations.

The guiding metaphor will be the analogy between foraging for rewards in a field and foraging for ideas. The domains differ, but the issue is the same—rummage through a space of "something valuable," picking those items most likely to return the most value to the rummager. The system already knew how to rummage for "good stuff," so it simply redefined what qualified as "good stuff."

The reward-prediction error signal described for the dopamine system is well-suited to choosing goals and to guiding sequences of decisions. The difficulty with pursuing goals is that they often require complex sequences of actions to obtain. The capacity to guide a sequence of actions using internally represented goals is called cognitive control. There are numerous books and papers on cognitive control and an equal number of theories, each emphasizing different, but important, features. Despite their differences, all these accounts implicate the prefrontal cortex as the site where goals are actively selected, maintained through time, and used to guide sequences of choices. One of the best ideas along these lines has been championed by a group of quantitative psychologists, Randy O'Reilly, Jon Cohen, and Todd Braver. They call their idea the dopamine gating hypothesis.

The human brain shares structural and functional similarities with those of other primates; however, the human prefrontal cortex is vastly expanded compared to all other primates. The prefrontal cortex is also a region of extreme information integration. It's like the ultimate information technology (IT) center, collecting parallel data streams from literally every part of the brain and providing direct feedback connections to *almost* every region. There are even direct connections straight back to the midbrain dopamine neurons and to the striatum, a collection of brain structures involved in the selection and sequencing of actions. We'll return to these connections shortly. Like the striatum, the prefrontal cortex is known to be involved in the selection and stabilization of goals. Currently, our best idea is that goals are represented by recurrent patterns of neural activity that are stable for some time and can be distinguished from ongoing background activity. Research over the last fifty years has contributed to our dramatically improved current picture of prefrontal cortex function.

Much of our knowledge of the role of the prefrontal cortex in selecting and sustaining goals comes from cases where this function has broken down due to injury or disease. A beautiful account of how behavior changes when the prefrontal cortex is damaged can be found in neurologist and neuroscientist Antonio Damasio's *Descartes' Error*, which chronicles the travails of Phineas Gage, a nineteenth-century railway supervisor whose ventral prefrontal cortex was obliterated in an accident. With this loss, Gage's personality changed in a variety of ways, the most dramatic being his incapacity to make sensible decisions. One area of Damasio's work has focused on patients with damage to their prefrontal cortex similar to Gage's. Two major themes have emerged. First, the ventromedial prefrontal cortex is a site where rational and emotional valuations are integrated. Second, this region, probably through the integration of different value streams, is intimately involved in goal-dependent decision-making. Patients with lesions to this part of the prefrontal cortex often display what is best described as pathological indecision, possibly due to absent or flat valuations across all their available behavioral options.

To pursue a goal, it must first be stable through time, that is, it must be held in mind. The prefrontal cortex is essential for this function, a fact highlighted by neurologists Donald Stuss and Robert Knight in a recent book. They detail disturbances of the prefrontal cortex that lead to chronic perseveration (fixation on a goal) as well as pathological distractibility (incapacity to keep a goal in mind). They even present cases where both are present in the same individual. The issue illuminated by such pathological states is the need for mechanisms that select an appropriate goal, sustain it long enough for it to be pursued, and dismiss active goals when they are no longer needed or a better goal is detected. The physiologist Earl Miller and psychologist Jon Cohen have built these data into an integrated framework for explaining these functions of the prefrontal cortex, and remarkably, reward prediction plays a central role in their story. I will marry their account to a mechanism that I believe helps to explain in computational terms one of the sources of creativity and flexibility in the human brain. It's here that I think theoretical computer scientists will be able to help the medical community in the near future.

IT IS NOW THOUGHT that goals are represented in the prefrontal cortex by stable patterns of neural activity. Despite the widespread output connections from the prefrontal cortex to other brain regions, the vast number of synaptic connections made in the prefrontal cortex start and end within the prefrontal cortex. It's mainly talking to itself! These self-connections are the primary cause of patterns of neural activity that can form here, a theme known in the neural network community as a recurrent network. A good metaphor for stable activity patterns in a recurrent network is fish schooling. Schools of fish show coherent overall motion and often resemble some great composite sea creature, undulating, reaching, reversing directions, like a giant sea amoeba. If a predator approaches, the school can swerve as a whole, adding to the impression that it's a single, unified entity. But take a closer look inside the school and there is a lot of local craziness going on, with fish darting in all directions. The overall motion is coherent enough to be called a pattern: a school. The schooling behavior is much like reverberating neural activity—all the "self-talk" between prefrontal neurons. In any circuit, self-connections cause debilitating pathological behavior. Just think of what happens with simple feedback through a microphone; it's uncontrollable and so not very useful. In the prefrontal cortex, there are specific mechanisms that encourage recurrent activity to settle into stable but short-lived patterns. We will skip the complicated ways that recurrent networks are made stable and focus primarily on how information from other brain regions selects stable recurrent patterns in prefrontal circuits. Here's what we know.

Extremely diverse information streams impinge on the prefrontal cortex, but they don't all get in. They are blocked somehow. It's as though this brain region had an *information shield* that it can turn on and off at will, like a Star Wars defensive screen for incoming information. Some of the time, the prefrontal cortex is shielded and talks mainly to itself, one idea chaining to another, and then sends off the results to other brain regions. At other times, it's opened up to information flowing in from all other parts of the brain. The mechanics of information shielding are not understood completely, but it's very clear that dopamine plays a central role in blocking and admitting information from outside the prefrontal cortex (turning the shield on and off), and stabilizing patterns of neural activity that have formed there (selecting a goal). So here's the job accomplished in the prefrontal cortex written as a stepwise recipe:

(1) Turn information shield on or off ("gate" information flow into prefrontal cortex).
(2) Select one pattern of neural activity from among several competing patterns.
(3) Stabilize selected pattern so that it can act as a goal.
(4) Use goal as a feedback signal to control other brain regions.

This cycle of gating, goal selection, goal stabilization, and goal feedback is typically called cognitive control. In our brains, it appears to be a relatively low-bandwidth function. We just don't think very fast, and when we do think, it's hard—especially if the ideas are new. Of course, following our efficient-computation prescriptions from before, we most likely think as slowly as we can, consistent with our survival.

In an attempt to understand how these steps could be carried out by the prefrontal cortex, the parents of the information-shield idea (O'Reilly, Cohen, and Braver; OCB) built a series of computer models to demonstrate how prefrontal goals could be selected, stabilized, and used to exert control—cognitive control—over processes in other brain regions. Like all reasonable scientific accounts, their story has a number of caveats and equivocations. However, they discovered something remarkable; they, too, had an "Asimov moment." Their model required a signal just like the reward-prediction error signal emitted by dopamine neurons in the midbrain. Why is this remarkable? Because their model was learning about which goals to choose, how to stabilize them, and how to learn to associate good goals with the contexts in which they were useful—all critical functions for what we would consider intelligent cognition.

In their early attempts, they conceived of the signal that did this job as a generic "gating" signal that (1) controlled information flow into the prefrontal cortex (the gate), and (2) stabilized the current activity pattern there (select-sustain a goal). At some point, they realized that all the properties they required for this generic gating signal were shared with the midbrain dopamine signal: the reward-prediction error signal from the last chapter. Neurophysiological data already suggested dopamine as the candidate for implementing OCB's proposal. Two signals possessing the same properties and both abstract signals requiring properties known to be associated with the neurotransmitter dopamine—it's reasonable in an efficient system to suspect that they are the same signal. Thus was born the *dopamine gating hypothesis*. Ironically, Jeremy Seamans, Terry Sejnowski, and Daniel Durstewitz were studying properties of prefrontal cortex neurons and came to the exact same conclusions on physiological grounds.

WE ARE NOW POSITIONED to marry the reward-prediction error function of dopamine neurons with the dopamine gating hypothesis. The former is the signal needed to pursue rewards. The latter guides the pursuit of something more abstract—a representation, a way to frame a problem. This is where Kasparov's brain exceeds Deep Blue; his brain must possess extremely efficient representations of the problems presented by chess games. Otherwise, his brain would not have been able to play Deep Blue to a near-tie. The proposal by O'Reilly, Cohen, and Braver exposed a remarkable instance of redeployment of reward-seeking mechanisms—the brain uses this same machinery to seek rewards in a field and "ways to represent learning problems." A magnificent reuse of a preexisting function. And as we discussed in Chapter 3, finding a better representation for a problem is exactly what reinforcement learning machinery needs in order to be efficient. Finding new ways to represent problems is a precise definition of cognitive innovation—creative thinking. Who would have guessed that the same signal that helps an animal navigate through a field to find food would be redeployed to help the prefrontal cortex find a correct abstract representation of a learning problem? And the connection to formal theories of reinforcement learning means that scientists are now in a position to understand some of the most important neural pathologies that plague our species.

These ideas for dopamine shed light on the strange symptoms that humans display when dopamine function is perturbed by disease or injury. Since both ideas can be studied as systems of equations, I am quite hopeful that new progress on diseases that affect dopamine function will be forthcoming. How exactly does this marriage work?

Reinforcement learning theory has established that a *reward-prediction error signal* like the dopamine signal is ideally suited to direct dual functions: action selection and learning. This is true provided that the learning system knows how to categorize the problem profitably and is not starting from scratch. There is now mounting evidence that the dopamine signal is indeed used in this dual capacity in the mammalian brain. By analogy, the dopamine gating hypothesis proposes that dopamine bursts in the prefrontal cortex direct two similar functions related to the goals represented there: (1) goal selection (similar to action selection), and (2) learning about goals. The dopamine bursts produced by the midbrain dopamine neurons are ideally suited to direct these two functions. Why? Because they can answer the following three questions:

Q. When is a goal update needed?

A. When something better comes along.

Q. How can we know that something better might have arrived?

A. Signal the presence of a stimulus that predicts future reward.

Q. Why does a reward-predicting stimulus warrant a possible goal change?

A. The predicted reward might be better than the promise of the current goal.

Earl Miller and Jon Cohen illustrate the moral of this story with a revealing example. You are walking to work. The sight of a twenty-dollar bill on the sidewalk catches your eye. A twenty-dollar bill is a reward-predicting stimulus, but only if you update your current goal of walking to work, and change it momentarily to a new goal—"Bend down and get the twenty-dollar bill."

The mechanistic idea is that the dopamine burst takes down the information shield of the prefrontal cortex ("opens the gate") and lets the current information establish a new goal state. There is clearly some kind of competition process that goes on at this stage, pitting the current goal against the potential new goal, but this process is not understood. Dopamine bursts occur either for unanticipated reward-predicting stimuli or for some intrinsically rewarding event. This is exactly the right kind of signal to use to take down the information shield and let the prefrontal cortex consider another goal. It tells the prefrontal cortex that "something better than expected may be coming along" and sets the prefrontal cortex looking for a new goal or new representation.

What about the learning function of the dopamine signal in the prefrontal cortex? What exactly would get learned at the level of the prefrontal cortex under the direction of the reward-prediction error signal? O'Reilly, Cohen, and Braver propose a straightforward but powerful analogy with the reward-prediction error model. Recall from before that the dopamine-prediction error signal acts to transfer value from intrinsically rewarding stimuli like sex and food to cues like images of food and sex that predict the future occurrence of the "real" rewards of actually obtaining food and sex. We also should remember that this value-passing scheme is arbitrary—another stimulus, like a song, that consistently predicts the images that predict the food will also accrue value. So a friend of my friend is my friend. Similarly, a predictor of a predictor of reward predicts reward. The dopamine system starts this sort of sequence by first giving a burst ("better than expected" response) to the food itself, and later, provided that the food is always preceded by the song and the image, the burst response transfers to the song—the earliest consistent predictor of the real reward. From the point of view of the dopamine system, the song, whenever it occurs, is a value proxy for the food that will be available in the near future. The idea of these three psychologists is that the new goal, admitted to the prefrontal cortex by a dopamine burst, will be the active goal at a time that the dopamine burst directs the rest of the brain to learn about what caused the burst—the "something better than expected" that just happened. Their idea for the learning part of the dopamine gating hypothesis can be presented as a Q-and-A. Suppose that a light caused the dopamine burst.

Q. Why is the dopamine burst happening (why a "better than expected" signal)?

A. An unexpected stimulus (the light) that predicts reward has just occurred.

Q. What happens to the new goal state just admitted to the prefrontal cortex by the dopamine burst?

A. Just like the "light predicts juice" learning, the brain learns "light predicts goal representation."

Q. So what?

A. A reward-predicting stimulus (the light) selects and admits a goal into the prefrontal cortex. This goal directs acquisition of the rewards predicted by the light. This just makes good sense—learn to select those goals that lead to reward.

Their proposal is excellent because it is a scheme whereby a goal gets selected (made active) by reward-predicting stimuli and helps the brain acquire the predicted reward by changing behavior. The specifics of their proposal are now open to a number of detailed experimental tests, but the idea passes the "sounds good, smells good" test. This idea is only the tip of an iceberg and suggests a mechanistic answer to the neural origins of the superpower. But the superpower goes one more step. Goal states like "going to the next level" or "I want to be a good person" don't just get associated with reward-predicting stimuli, but come to act directly as rewards themselves—it's like they plug right into a slot in our mental software labeled "reward." What are the consequences?

THE IDEA SIMPLY SOUNDS dangerous, doesn't it? Arbitrary goals plugged into reward sockets, like plugging the output into the input—almost always a bad idea. The reason for the danger: Well, they aren't called "input" and "output" by accident, and besides, it has a kind of "fork in wall socket" ring to it. This could be very dangerous. Most systems are not designed for such recursion, but sometimes recursion can be good, even creative.

The reward prediction machinery in our brains provides a way to understand the selection and stabilization of goals represented in the prefrontal cortex. The reward-bootstrap idea, embodied in the dopamine gating hypothesis, provided a mechanism for this goal-selection process. But there are some important missing items. First, a goal can't last forever, since it's only likely to be useful in a specific setting and for a limited amount of time. There must be active mechanisms for removing goals from the prefrontal cortex. Indeed, we know there are basic cellular and molecular mechanisms available to prefrontal neurons that limit the lifetime of any goal representation. There is also evidence that the striatum is involved in goal selection and maintenance.

No matter what the mechanistic explanation at the biological level, goals must be evanescent and actively habituated; there must be mechanisms to remove a goal state from the "active" list. Otherwise, we could become locked into the pursuit of a single goal, a feature not associated with survival. This "active-list removal" feature should be true for any goal, even the really important ones like chasing food, water, and sex; however, these goals are special.

The prefrontal cortex must possess a "let it go" mechanism that displaces goals from the active list if they have been there too long, or if they prove to be unhelpful, or if they are downgraded by other, higher-value goals. However, there are special classes of goals that have higher priority almost always. Goals that must be pursued to keep the species going—like food, water, and sex. And although we will habituate to these goals for a while, our brains will eventually elevate them again to high priority. No matter how much you eat or drink now, you will be hungry again later when your energy reserves run down. No matter how much sex you have now, it will be put back on the active list as a goal later on. The reason for having a few special goal categories is that all other goals that we might pursue must first "fit in" with the basic needs of survival and reproduction. Just like the Hippocratic oath implicit in the evolution of new genes—"first do no harm." Genetic innovation must first "fit in" with all the other genes that have kept an organism surviving so far. Innovations that fail to fit in, that mess up something essential for survival, don't last long, if at all. One way to meet this requirement and not be eliminated quickly is to be neutral—one central concept in the way we think genes change through time. Likewise, there is an implicit "do no harm" ethic constraining the ideas that can act as goals and forcing the new ideas to "fit in" with our special goals of food, water, and sex.

Until recently, an idea that vetoed the reward status of sexual intercourse during fertile years was incompatible with reproductive success. Modern technologies may eliminate this constraint in the coming years, but for now it's fairly safe to claim that an idea that completely vetoes sexual intercourse would not have survived for too long. If sex does not maintain its reward status throughout the fertile years of a young man or woman, then the brain could not ensure that reproduction would likely occur. This leads to the conclusion that "idea" goals must acquire less priority when they conflict with food, water, and sex, putting these goals in a special category. So there must be special reward categories that always represent something valuable from the nervous system's perspective.

These special rewards act like other stimuli to your nervous system, represented by your prefrontal cortex, but possessing a sort of command status—the organism must accede to their wishes in order to survive and propagate. This is why I'm arguing that there must be special reward categories for the brain. Behavior teaches us that things can't be this rigid. Once someone is sated on sex or food, there is no reason right away for sex and food to have privileged status as a goal. The point that I am making is that the capacity of the brain to learn value proxies for food and sex, and therefore learn to predict their likely future occurrence, should never prevent them from acting as goals. Consequently, these survival-dependent "rewards" hold a special status in our nervous systems. And this observation is what I think forms the basis of the superpower's capacity to veto instincts. What would happen if the system found a way to put some "idea" reward into the category of sexual intercourse? That is, not just let the abstraction act like a reward, which we think happens in order that we might pursue ideas. Instead, the abstraction goes into this special category of reward—one that always receives priority when there are conflicting goals.

Let's repeat my thesis on the biological origins of the superpower as a kind of recursive trick discovered by the mammalian brain and existing in its most developed form in humans. Plug prefrontal goals directly into the "special status reward socket." In this way, ideas generated by your prefrontal cortex act directly as high-priority reward signals from the point of view of the prediction error systems. These ideas then act directly as the reward signal to the dopamine neurons, which then try to combine information from other regions of the brain to predict this new "idea reward."

The proposal here goes one step beyond the gating hypothesis above. Instead of merely associating goals with reward-predicting stimuli, goals become rewards. And in some cases, goals may become "high-status rewards," in the same way that food and sex are high-status rewards. Both cases yield something like a behavioral superpower—capable of vetoing other goals that we might have. How might all this feel to us? I have not been discussing feelings, only the covert computational signals that underlie them. But personal experience teaches us that ideas, especially those that gain reward status, really can "feel" like something. They engender feelings typically associated with rewards—a kind of mild, satisfying yet fleeting, euphoria.

When the theorem is proved, or the house built, or the painting completed, or the music interpreted, or the phrase uttered—when some sought-after abstract goal is experienced—it feels almost like a drug, or maybe more like a fainter version of the satisfaction from food or sex. Nevertheless, it really feels like something. Of course painting, architecture, music, poetry, and mathematics are not primary rewards. They are not required to survive and reproduce, but they can feel that way and we act so as to give them meaning. We will often rearrange our lives to pursue them as though our lives depended on them. And for those of us not inclined to be artists or mathematicians, there is something similar that hits our personal sweet spot. And yet these things don't feed us or give us more progeny. How did these culturally dependent goals gain such value? The goals became rewards. One can see how this mechanism could ruin an individual's chance of reproducing, and that's why there must be a lot of control over this process if such a mechanism is to fit in with the requirements of survival.

The system found a way to take highly integrative states (goals) in your prefrontal cortex (let's think of them as ideas) and plug them into the reward socket that feeds reward information to your dopamine system. As if by magic, the goals act as the current reward signal! This is truly a remarkable trick and its implications are still emerging. The dopamine system doesn't know the nature of this signal. It doesn't need to. The dopamine system simply combines information from other brain regions with this reward signal, produces a reward prediction error, broadcasts it widely, and teaches the rest of the brain about the new reward. These "idea rewards" then pass their value to other brain states that predict them just as lights that predict juice gain a kind of value-proxy status, representing the future promised value of the juice they consistently predict. This is a recursion trick. Plug the goal state in directly as a reward. The protection from pathology here arises because goal states don't last forever; they decay away.

The recursion trick is not unique to humans—what is unique to humans is their prefrontal cortex and their ability to represent the world in a multitude of ways and to remember these representations. The combination of prefrontal cortex function and the recursion trick may have been a crucial step that conferred real social agency to human behavior. But notice that this agency is not magic. Our understanding of it can now be cast in physical terms and parts of it cast as mathematical problems. I'm not selling social agency from some murky, nonphysical place; I'm offering proximate mechanisms as part of the physical substrate of social agency.

For this recursion trick to work, a lot of control must also exist. There must be some kind of natural limitation on two features: (1) the kinds of goal states that can replug into the dopamine system to act as rewards, and (2) the lifetime of the effect. This suggests that the kinds of goals that can be conceived may be prefiltered; that is, we can't have just any idea and have it act like a reward. Instead, possible goal states may be treated to some kind of censoring function before becoming available to plug into the reward prediction machinery. Alternatively, the reward prediction machinery itself may exercise selective control over which prefrontal goals (ideas) it will treat as rewards. This latter possibility would almost certainly involve the striatum, a region known to possess an anatomy and physiology that would support this idea. But the basic constraint is clear. The capacity for an idea to "plug back in" as a reward signal is such a powerful capacity that it must be subjected to numerous levels of control. It's my feeling that this is one of the weak points in our decision-making system's

behavior, and it's here that I think that the computational models offer new ways to view important neurological and psychiatric diseases.
Let's explore them.

WE ARE BEGINNING to obtain real scientific insights into the way
that we have agency over our behavior—willful, idea-based control.
This same perspective shows us how social constructs can have a real
neural punch, how cultural constructs acquire a neural reality. The
actions of the Heaven's Gate cult provide just one example. Humans
can and will die for ideas—ideas about social equality, ideas about
laws, ideas about self-determination, ideas about religion. Current
events make all these familiar choices very real. The superpower argument goes some distance in suggesting a physical explanation for
how ideas gain this kind of power; however, it does not prescribe a
principle for choosing one idea over another. It's mute on how ideas
compete for status as rewards; however, the question is clear and I
suspect that we will see new work on these issues in the near future.

Let's not misunderstand this idea of the superpower. It cannot
possibly be infinitely flexible in its currently evolved form. We should
not imagine that a particular individual can by some force of will
make *any idea* gain the power of a primary reward and then overdose
on that idea—not willfully. But idea addictions do indeed exist—just
ask anyone who has a relative with obsessive compulsive disorder,
body dysmorphic disorder, or any other form of obsessive thinking.
It's quite possible that these are exactly idea addictions. And let's be
frank—these strange disorders can be debilitating, just as an idea
overdose might be.

The value of the computational perspective is that it produces a

mathematically tractable way of understanding the relationship between brain and mind, and it has provided some of the first fragile insights into how we value the world around us and choose based on
valuation. Computational models of brain function are still in their
infancy, a fact that also holds for value-based models of mental function presented here; however, they possess advantages that verbal descriptions lack. One advantage is that they quantify patterns of neural
activity and behavior in ways that make it easier to see the relationship between the two. They also provide a new way to categorize the
way that valuation systems break and the kinds of dysfunction that
would ensue.

We can address a number of common diseases from the point of
view of reinforcement learning models. My discussion is necessarily
shallow, since each disease represents a serious affliction for which
an enormous amount of biological and behavioral work has already
been carried out. It's not my hope to explain any of these diseases, but
to point out that some of them have provocative connections to some
of the computational models presented above. Maybe a clinician may
see a symptom in a new light.

Inspired by the reward-prediction error model of the dopamine
system, David Redish used a modified version of it to address some of
the essential features of drug addiction. Redish, known for his work
on the hippocampus, a brain region critically involved in learning
and memory, may have uncorked an important bottle. Addiction is a
complicated biological process that influences neurobiology at many
levels—neurons, synapses, membrane receptors, intracellular biochemical networks, and gene expression. Addiction is also a complicated set of behavioral phenomena that includes habit formation,
aberrant goal-directed behavior, and other fundamental and often

sweeping changes in cognition. People don't become addicts in a vacuum, and so there is an entire social psychology story behind most
facets of addictive behavior. The causes of addiction are really more
like a collection of susceptibility factors that divide naturally along
the three streams of influences that I have just described—biology,
psychology, and socioeconomics. Descriptions of addiction at all
these levels contribute to our current understanding of addiction and
will continue to contribute in the future.

There are lots of published "theories of addiction" that emphasize
only one or two of the streams without a detailed consideration of the
others. And this is to be expected. These are broad and complex areas
of knowledge that are hard to fuse in a facile way. However, I think
that Redish's translation of several important properties of addiction
into computational terms opens the door for better cross-talk between these areas and for testing new ideas suggested by the equations. In essence, Redish has turned addiction into a neural numbers
game and connected it back to theoretical descriptions of learning
and memory. This view is supported by recent experiments in rats
that also impinge on this model-based approach.

BEFORE RECOUNTING REDISH'S MODEL, I think that this is an
excellent point at which to review the origins of the reward-prediction
error model for dopamine, since it provides an example of why it
pays to pay for research. Let's face it, at this point, some readers may
wonder where all these theoretical ideas originate. Certainly my relatives wonder about this. For a number of years (basically until I was
thirty), they wondered exactly what I was doing and who was paying
me to do it. The argument went something like: Why should someone

pay good money just to hear you go on about things in exactly the way you did when you were a kid? I'm convinced that there are sympathetic heads nodding while reading this because of something analogous said to them. Unless someone grows up around science (which I didn't), it's really hard to envision exactly how it operates because when it's described in real terms (not glowing recollections), it just doesn't sound like a feasible way to make a living. It's always hard to know when progress is being made, the job description is often nebulous, competition for funds is fierce, and yet it maintains a visceral attraction to some (including me).

Many of the theoretical ideas presented in this book originated or were developed in special groups around the globe dedicated to theoretical computer science, theoretical neuroscience, or cognitive science. These groups employ a rich collection of different kinds of intellects, some with backgrounds in physics and math, some with backgrounds in psychology and cognitive science, and some with backgrounds in philosophy, engineering, and various other disciplines. Such collections are basically paid to invent the future of fields and, in this case, the coming computational revolution in the brain sciences. Despite the presence of early pioneers in the field, this computational revolution is only now fully under way. Twenty years ago, there were only a handful of places where theoretically minded neuroscientists could actually have a career and pursue work in the midst of a critical mass of others similarly engaged. Let's rewind to the early 1990s.

The earliest application of a reward-prediction error model to dopamine neurons was proposed in 1991 by Peter Dayan and me while we were working at the Computational Neurobiology Lab (CNL) of Terry Sejnowski. Terry, a physicist by initial training, has provided a valuable and unique resource to the neuroscience community for the last eighteen years. His group is housed at the Salk Institute for Biological Studies, a private science institute in La Jolla, California. Everyone who has ever worked at "the Salk" (as it's often called) knows that it's a special place, a place where everyone feels emboldened to reach for the next great thing, the next step in understanding. That same spirit of living at the "edge of the possible" still pervades the Salk today. The list of notables associated with the Salk reads like a who's who in biology, and even the writer Michael Crichton passed through the Salk Institute in the late 1960s as a postdoctoral fellow. If you are interested in cutting-edge biology, this is one place where it happens. And in the early 1990s, if you were interested in how the brain computes, the Computational Neurobiology Lab was the place to be. It remains near the center of the computational neuroscience galaxy, but the effort has taken root in some form in almost every major research operation in the world.

The stated purpose of the Computational Neurobiology Lab is to understand the computational resources of brains—period. The focus of all the work in the lab is to understand everything the brain does in terms of computation—biology, especially neurobiology, completely decomposed in terms of information processing. This ranges from the computations that molecules carry out to the computations carried out by whole brains. This vision attracts an eclectic and maverick group of scientists and may today sound old hat, but at the time, only MIT and Caltech had coherent groups with critical mass dedicated to computational neuroscience.

The computational neuroscience group at Caltech rallied early on around the work of John Hopfield and Carver Mead, and is now led by the creative efforts of Christof Koch and others. The group at MIT was organized early around the work of Tomaso Poggio, who is a leading figure in the computational understanding of biological intelligence. There were, of course, individuals scattered around the world with related interests, but these three groups formed the real triumvirate for computational neuroscience in the early 1990s, and they have produced an array of talented scientists who now populate institutions around the world. Computational neuroscience has now grown into a professional, worldwide effort with spin-off subfields, textbooks, and annual conferences. Lots of people are interested in how biology computes. My interest at the time was to use theoretical methods from the field of neural networks and machine learning to understand how the brain self-organizes its structure and its learning algorithms. At that time, the only problem was that I didn't know anything about machine learning; but fortunately, Peter Dayan did and was quite willing to introduce me to the area—flog me with it, actually. Dayan, at the time a freshly minted Ph.D. from the University of Edinburgh and formerly of Cambridge U., possessed an information-processing rate and generosity rivaled by few. The resources did not end with Peter; the lab was a literal hothouse of efforts to apply neural networks and machine-learning theories to real brains made of real neurons. "How does the brain work?" was our mantra, but nobody really knew how to proceed. Terry's solution was very much that of a stew-maker—take one of these, one of those, a little bit of this, and a dash of that. Mix thoroughly. Bam! "Testable and useful theories of brain function; experiments to follow." His ingredients were people—mainly postdoctoral fellows and graduate students—and his brewing pot was afternoon teas. This is where ideas were hatched and egos were broken. Terry's formula was to start tea with some kind of issue, spice it with the presence of some provocative visitor, and let the sparks fly. The spark production was catalyzed by the near daily presence of the late Francis Crick of DNA fame, neurophilosopher Patricia Churchland, and physiologist/neural theorist Chuck Stevens. The fact that many of the visitors were luminaries in their fields of study only enlivened things more. I and others continue to work on ideas and half-ideas spawned at those teas, and I have not seen the sheer energy of that gathering reproduced anywhere else. Despite the seemingly esoteric nature of the work carried out, practical consequences flow out of these kinds of places. This has historically been true in physics and it's also true for biology. Let's turn to the practical application of reward signals. What exactly is Redish's model of addiction?

Let's consider again how the reward-prediction error signal, encoded by the time and duration of brief dopamine bursts, transfers value to reward-predicting stimuli. It's a disruption of this process by drugs of abuse that Redish's model associates with addiction. The reward-prediction error signal can be directly related to a family of models called temporal-difference reinforcement learning models (TDRL). These models use an error signal called the temporal-difference error that resembles almost exactly the dopamine signaling recorded in the brains of monkeys by Wolfram Schultz of Cambridge University and more recently Paul Glimcher and Hanna Bayer at New York University. The abstract learning problem that TDRL solves involves a subject (the agent) that moves around in a world with a finite number of states. This could be actual movement of its body or simulated movement—thinking about moving around. The subject moves around the world, the states change, and the agent gets feedback in the form of a critic signal, here called the TD-error. The goal of learning for a TDRL system is to employ this prediction error signal to learn the value of each state of the world, after which the value of each state is stored within the agent. One important feature of these values is that they are like a stored form of *longer-term judgments* about each state of the world, because they represent, in a single number, the long-term rewards expected from that state into the distant future. The issue of how to make these kinds of learning systems scale up to really hard problems is an exciting and active area of research. TDRL systems stop learning once the values of all states of the world are estimated correctly. This kind of learning system places a pressure on efficient representations of the outside world. Creatures with brains that can efficiently categorize the states of the world relevant for a task can use TDRL to learn rapidly the details of the task. Inefficient representations of learning problems can cause TDRL systems to grind to a halt and learn virtually nothing; there is a real premium on framing the problem correctly. Perhaps this is why goal selection and reward-prediction error signals are so intertwined.

The progression of learning in TDRL goes as follows. A stimulus that delivers reward, like a juice squirt in the mouth, causes a transient burst in dopamine neurons signaling the rest of the brain, "That was better than expected." Now suppose that a light consistently precedes this squirt of juice. Through multiple rounds of the experience—"light predicts juice about six seconds later," "light predicts juice about six seconds later," and so on—a TDRL system transfers the value associated with the receipt of juice to the earliest stimulus that predicts it, here the light. The light now has value, or more precisely it's a proxy for something of value that will arrive six seconds in the future.

At this point, an unanticipated onset of the light will cause a dopamine burst that means "better than expected," but only because it has consistently predicted a future event of real value (juice squirt).

This value-passing scheme is powerful and allows values from all kinds of complex interactions to transfer back in time to stimuli in the world that predict them. The trick is that once these values correctly anticipate future reward, the system stops learning—the values match reality and so they don't need modifying. Once the world changes a little, the values may need to be updated, and the dopamine bursts will signal this. For example, let's revisit the honeybee example from before. Suppose that I'm a bee and I have a model that blue flowers yield on average ten units of nectar and red flowers yield on average twenty units of nectar. I will spend a lot of time pursuing red flowers, but while I'm doing so another competitor may be eating the nectar from the blue flowers, making my model of their likely nectar yield wrong. When I finally decide to forage on the blue flowers, I will find that my nectar experience does not match my internal expectations, my reinforcement learning systems will produce robust error signals, and I will readjust my internal value functions to reflect the changes outside.

This is where Redish's model steps into the picture. His claim starts first with an assumption: TDRL models describe the simple ways that dopamine bursts and pauses act as reward-prediction error signals that guide the learning of internal values. Redish then claims that addiction is all a matter of mistaken identity in the form of false value. Keep the "light predicts juice" scenario in mind. Once the value assigned to the light predicts the time and reward value of juice, the TDRL system stops learning. Substances like cocaine never let this happen because they sidestep the checks and balances that allow the system to stop learning about natural rewards once they are adequately predicted. Instead, each cocaine hit continues to cause a dopamine burst signaling the system that "something with extra value has just arrived," which is just another way of saying, "Something better than expected has just arrived." For juice, the system eventually learns to stop signaling this because it learns the value of the juice, but since cocaine *always causes a little dopamine burst*, the TDRL system thinks that there is more value still to learn about. In Redish's interpretation, this means that the learned values of stimuli that predict the cocaine will never stop growing—they will get more and more valuable with each use of the drug. In practical terms, this would mean that drug paraphernalia and the drug delivery that they consistently predict would gain more and more value in the brain of the user. So the sights, sounds, and thoughts surrounding drug-taking would continue to grow in value and continue to get harder and harder to resist. And there is no theoretical reason why this would stop even if the drug had lost its euphoria-inducing quality.

This same TDRL model of learned value accounts for an influential psychological model of "wanting" proposed by Terry Robinson and Kent Berridge. Berridge and Robinson call their psychological model incentive salience. They have used their ideas about incentive salience to explain the behavioral difference between "liking" and "wanting" and they have shown how manipulations of the dopamine system can impair an animal's ability to learn "what to want." There is now an impressive catalog of behavioral data that can be explained within their framework.

Understanding the TDRL model's connection to "learned wanting" is an important step, since it makes the strong prediction that a dopamine fluctuation does not encode "pleasure," nor does it directly encode "wanting." Instead, wanting is related to the underlying value learned through the guidance of a dopamine-encoded reward-prediction error signal. So the "wanting" part is a parameter stored in

ervous system while the measured dopamine fluctua-
...senting prediction errors, reveal the underlying differences in learned value. Changes in dopamine reflect "differential
wanting" across two alternative choices. So the wants aren't measured
directly by the dopamine signal, but differences across wants are. This
marriage of the best traditions in experimental psychology to computational interpretations provides new insights into addiction. "Wants"
alone are stored value functions, like the stored valuations in Kasparov's brain or Deep Blue's program. "Differential wanting," differences across wants (relative preferences), shows up as reward-prediction
error signals encoded in dopamine fluctuations. With that in mind, let's
return to Redish's TDRL addiction model.

There are several reasons to take Redish's use of the reward-
prediction error model seriously. First, it provides quantitative predictions about how different drugs (like cocaine) could influence
learning and decision-making in a manner that can be simulated on a
computer and understood in computational terms. Secondly, it addresses concretely how the brain's valuation of a drug and the costs of
a drug should trade off with one another—that is, the model can address something as practical as voucher schemes to keep addicts from
taking drugs. Lastly, Redish's model is consistent with a remarkable
series of experiments by Mark Wightman and his colleagues at the
University of North Carolina. Wightman is an electrochemist who,
with the help of numerous collaborators over the years, has pioneered
new methods of making subsecond measurements of fast changes in
dopamine delivery to different parts of the brains of rodents while
they are freely moving about. This remarkable technology has measured dopamine pulses in brain structures known to be involved in
decision-making and goal-setting as an animal learns to associate a

light with the subsequent delivery of cocaine. This group has provided direct experimental support for the idea that a reward prediction error is encoded in dopamine release and that these brief squirts of dopamine can initiate drug-seeking behavior. The ability to relate Wightman's experiments to computational models has propelled this field in recent years, and there is much hope that this new quantitative understanding and its descendents might translate into principled therapeutic interventions in addicts.

These modeling efforts also equip us with a language to categorize neural diseases in a new way. For example, it allows us to see a natural connection between Parkinson's disease, normally thought of as a movement disorder, and addiction. David Redish's TDRL model of addiction to drugs of abuse shows how repeated use of a drug injects "false value" into the nervous system and keeps increasing the learned value of the drug. Through normal learning mechanisms present in the brain, this "false drug-value" is transferred to cues that predict drug-taking. This transfer means that the decision-making machinery in the brain will choose actions that lead to the cues which themselves lead to receipt of the drug, *which always delivers "extra" value*. This is an insidious cycle.

The valuation perspective highlights an important fact: The brain's decision-making machinery is acting rationally; the pathological computation is the *value* computed for drug-predicting cues, not the decision made based on that value. It's quite rational to choose the choice that returns the most value for the least behavioral investment. Drug addiction disturbs our ability to learn correctly that value, that is, "what we should want" or "how much we should care about drug-taking." What, then, does an addict's brain want? It wants to experience those images that predict subsequent drug use (they have learned value) and it wants the actual drug—it, too, has value because the dopamine burst that attends each drug hit reports that "there is unaccounted-for value associated with this drug." It's a tragic use of normal learning machinery.

The addict becomes addicted to the idea of the drug as well as to the drug itself. Addicts literally become obsessed with thoughts of the drug and the experiences that surround it, and a simple TDRL model explains this in terms of pathological value functions learned by the brain. This is an *idea addiction* due to pathological valuations. It's like messing around with Deep Blue's value function so that it always moves so as to protect the rooks—no matter what. If this pathology were present, Deep Blue would behave in inexplicable ways, frustrate its creators, and it certainly would not beat a champion chess player, who could easily exploit its "rook addiction." We would describe the program as being pathologically hooked on the idea of saving its rooks, but the program would see nothing wrong with that—deep in its heart, its values have been reprogrammed.

There are some natural social rewards that can also lead to similar and equally debilitating idea addictions in humans. Drive around any modern city and you will be inundated with images of how your skin should appear, what car you should buy, how small or straight your nose should be, how you should wash and dry your hair, and so on. The critical word here is *should*. These images transmit goals that the rest of us should seek. Typically, these goals garner extra behavior power because they are paired with sexual attraction, fecundity, masculinity, money, and, well, all those things that confer status in our species. Status, like food and water, has value and is therefore rewarding to humans. Link an image to status and the image acquires value. It's not really surprising that all sorts of disorders of valuation afflict modern-day humans. I want to address these disorders briefly with the TDRL model in mind.

THERE ARE A VARIETY of mental disorders where obsessive thoughts invade a person's mind and dominate his life. Sometimes these diseases sound odd and it is tempting to wonder, "Why doesn't he just choose not to do that?" but this attitude misses the absolute automatic nature of these diseases; people so afflicted really have no choice. Deep down their brains value things differently from a normal brain.

These diseases include obsessive compulsive disorder (OCD), body dysmorphic disorder (BDD), and other related thought disorders. OCD and BDD have the common feature of pathologically intrusive thoughts. In obsessive compulsive disorder, these thoughts recur involuntarily and frequently and are literally debilitating. Obsessions typically include some fear of the sufferer, like being injured or falling prey to a deadly disease. The compulsion part usually involves repeated, ritualistic acts used to ward off the obsessive thoughts— compulsive hand-washing to destroy disease-causing germs. Body dysmorphic disorder shows similar patterns, but related more specifically to some small imperfection in one's body. These imperfections are either real or imagined, and include minor deviations in face symmetry, the length of the nose, and disgust over minor skin blemishes. The list is long and the items particular and idiosyncratic. To rid themselves of these disturbed body images, subjects with body dysmorphic disorder will go through surgery or take on a grueling exercise regimen. These remedies and the obsessions over the imperfections cause tremendous anxiety and basically ruin one's day-to-day existence.

OCD and BDD are well-defined clinical designations with variability in the character and prevalence of the symptoms. I'm not going to offer some silly one-size-fits-all solution to them, but I wanted to mention the possibility that they, like substance abuse, may fall into the category of valuation-system diseases. We have been discussing how pathological valuations engendered by exogenous drug-taking can subvert the brain's decision machinery by sending it "fake valuations." In the TDRL model of addiction offered by Redish, the drugs send an anomalous value signal to the brain, but the response to this signal is intact; the brain learns correctly and makes correct decisions to pursue more valuable behavioral options over less valuable ones. It's very much like Sigmund Freud's view of the paranoid personality—only the premise is wrong. The rest of the paranoid person's behavior is quite rational. If you thought that there was an enemy behind every bush, you, too, might rationally follow extreme and desperate patterns of behavior.

In BDD, one possibility is that the sensitivity to value deviations is way too high—that is, the person is responding to real, but normally minor, deviations from perfection that we all possess, but with the "volume" turned way up. Once the brain learns incorrect values for these imperfections, the catastrophic decisions that follow are really quite rational. The same kind of problem may contribute to OCD: The thought gains too much value and so stimuli that predict the thought also gain value, and so on. The brain builds a network of false value proxies so that "all roads lead to Rome," the obsessive thought.

ONE LAST BUT IMPORTANT DISORDER that may have a broken-valuation component is Parkinson's disease. Parkinson's disease is a neurodegenerative disorder where eighty to ninety percent of the dopamine neurons in the substantia nigra die and what is described as a progressive movement-disorder results. One main target of these dopamine neurons is the striatum, a set of brain structures involved in the selection and sequencing of actions. This simple view of the striatum has been changing recently. The striatum is now known to be involved more intimately in selecting and pursuing goals, and in the processing of reward-related information. Patients with Parkinson's disease develop emotionally flat expressions on their face and have an escalating difficulty in initiating movements. It is common that Parkinson's disease is accompanied by depression and it's known to be associated at younger ages with brain injury, particularly those that occur in boxing. The movement problems and stiffness associated with Parkinson's disease are familiar to most people, but it's less well known that there is a range of cognitive problems, including memory loss, impaired visual function, and changes in emotional processing. Once symptoms of the disease are prominent, most (more than eighty percent) of the dopamine neurons have already been lost.

One singular result suggests that pathological valuation contributes to this disorder—more specifically, a lack of differential valuation. Early in the disease, patients will be brought to their physician, extremely stiff and almost incapable of initiating movement. The patients will be given a drug that transports into the brain and is converted into dopamine, and by the afternoon, this formerly stiff patient is walking about and vacuuming the living room floor. I'm exaggerating to make my point. For many scientists and physicians, these are troubling data. Wash a Parkinson's-afflicted brain with dopamine to change baseline levels in the brain and, bingo, you restore the ability to initiate a correct and long series of complex actions. The "dopamine wash" could not have been directly involved in the computation of the movements that were carried out, but must have been permissive in some important way. What's possibly going on in terms of valuation?

A novel contributing factor to the Parkinson's disease (PD) problem may be something like pathological indifference to available actions. And this indifference results in pathological indecision. I don't mean either of these in a conscious sense; instead, I mean that the system for valuing possible actions is no longer able to differentiate the value of one action over another. It's a classic signal-to-noise problem. In our current model-based description, the difference in value between two behavioral options is "read out" by dopamine fluctuations. But in PD, these fluctuations have become so small that they are all about the same size; technically, they are buried in the noise. This is terrible because now the brain structures that "read" the dopamine fluctuations cannot discern the difference in value across the available possible outputs (actions). There is no "differential value" in choosing one action over another. So the decision machinery in PD brains sees a flat value function for all possible actions, and rationally decides to stay in the current state; hence, the PD subjects *actively freeze*. PD freezes the decision-making machinery by presenting it with a flat value function across choices. Freezing in this case is completely rational and efficient. The mechanism asks itself: "Why waste energy making a transition (movement) to another state when that state is exactly the same value as my current state? Instead, just freeze." A rational freezing response is revealed by thinking about the underlying valuation functions.

This same idea was highlighted in the excellent and almost lyrical account of the flat valuation problem by physiological psychologist

Peter Shizgal. In recounting the exact problem of pathologi
ference he wrote:

> In his nihilistic novel, *The End of the Road,* John Barth in-
> troduces a protagonist who falls prisoner to his own indeci-
> sion. To Jacob Horner, the predicted consequences of all
> actions have become equivalent. Thus, he sits, immobile, on a
> railroad station bench. Frozen there throughout an entire
> night, Horner is eventually rescued by an unorthodox thera-
> pist who shakes him free of his paralysis and provides him
> with three rules designed to prevent a recurrence. "Sinistral-
> ity" selects the leftmost of a set of items, "antecedence" selects
> the item first encountered, and "alphabetic priority" selects
> the item whose name appears first in an alphabetically sorted
> list. By applying these rules sequentially, Horner will be able
> to choose a course of action in a wide set of circumstances.
> Given his professed indifference to consequence, it need not
> matter to him where the application of these arbitrary choice
> rules leads.

> Barth's deft portrait of anomie stands in sharp contrast to
> the Darwinian absolutism of the natural world, where selec-
> tion has imposed an ultimate goal: making grandchildren.
> Unlike Jacob Horner, animals in the natural world behave as
> if they are anything but indifferent to the consequences of
> their actions. Their choices are guided, not by whimsical
> rules, but rather by systematic assessment of the current
> stream of sensory input, of the recorded consequences of past
> actions, and of physiological and ecological conditions. The
> challenge for the cognitive and behavioral neurosciences is to

understand the processes of evaluation and decision and to uncover how they are implemented in the brain.

In Shizgal's account of Jacob Horner's problem, all options have the same value, and consequently the decision-making mechanisms can't decide which action to choose. Here's the logic from a first-person perspective:

(1) All options are equally valuable, including remaining in my current state.
(2) Making a transition to another state requires energy.
(3) No other state is more valuable than the current state; therefore, command the system to stay put—freeze.

It's not that they can't move—their decision machinery actively chooses not to move. In this scenario, *stiffness is an efficient decision* because all possible actions have been estimated to have equal value, so the system makes sure that it stays in the current state. Why should the Parkinson brain view all actions as equal-valued? This possibility is probably not true. Instead, the Parkinson's brain probably retains the correct stored values of different actions. The problem is that dopamine fluctuations have become very small and very noisy. They are normally used to "tell the difference in value" of possible actions, but in Parkinson's disease these tiny dopamine fluctuations cannot discriminate adequately the value of one action over another. So the downstream decision-making mechanisms that sense the dopamine fluctuations "see" that all actions have roughly the same value, and they make the efficient decision: "Freeze—don't waste energy on actions that yield no extra value to the organism."

Why should washing the brain with dopamine alleviate this rational stiffness problem? It's possible that increasing the dopamine level in the remaining ten percent of dopamine neurons increases the size of the dopamine fluctuations and temporarily restores the capacity to differentiate the value across the available actions. In this case, the downstream decision-making machinery can select an action that promises more value than the current state, and so movement sequences occur. All the movement sequences are precomputed and ready to go as needed; restoring the dopamine fluctuation lets the system choose the sequence with the highest value. Neuroscientist and psychologist David Eagleman developed this idea in his Ph.D. thesis using a TDRL model exactly like that employed by Redish. The importance of his effort is that it provides a new way to conceptualize this terrible neurological disease—a decision-making problem subsequent to the loss of the differential value signal—here encoded by dopamine. Parkinson's disease has many features not captured in this description, but the learning models have helped produce a new hypothesis about some of the important and debilitating symptoms.

THE FIRST FEATURE OF GAMBLING games to make clear is that they evolved around humans to exploit their valuation and decision-making machinery. The fact that humans are terrible at playing gambling games rationally testifies to the games' capacity to slip by our normal decision-making defenses. If you don't believe this assertion, just go to any major hotel in Las Vegas, look up, and count the floors. Who is losing? Well, I can say for sure that the gambling games are not losing. What could be more irrational than feeding money into a slot machine without any knowledge of the rate of return for pulling on the handle or pressing the start button? Nevertheless, humans love to play slots, and some even think that they are skilled at it! We will address a market investment game in the next chapter to get insight into how gambling games sidestep the natural checks and balances of our valuation and decision machinery. These games lull our nervous system into a kind of dull complacency and exploit the fact that we don't know how to think about the probabilities involved. We over-represent rare events and we are suckers for seeing patterns where no real statistical patterns exist. It's no wonder that we see the cartoon character Snoopy in cloud formations.

One of our worst capacities involves the detection of statistical streaks—winning streaks, losing streaks, "hot" streaks, "cold" streaks. Humans are notorious for inferring a real streak where none exists. The most revealing example of "ghost streak" detection was exposed by the late Amos Tversky, well known for his pioneering work on decision-making with longtime collaborator Nobel Laureate Daniel Kahneman. We will return to Kahneman in a later chapter, but for now let's hear about Tversky's *hot hand* analysis of basketball shooting streaks. Flurries of popular press articles surround the results of the work, but one fact is certain: Humans are terrible at correctly perceiving streaks of successful shots. Studying films from well-known basketball teams, Tversky and his colleagues showed that neither the length nor the frequency of shooting streaks exceeded that expected at random. Here, *at random* means that each shot is independent, and on each shot, the player "shoots his average." Controversy notwithstanding, Tversky was right. Our issue is that such faulty streak-detection is exactly the kind of foible that a gambling game easily exploits. After a number of "wins" on a slot machine, we may continue to play, lose money, and eventually quit, not realizing that the house has taken a little bit from us. Average this process over thousands of generations of human brains, and pretty soon you have a fancy fifty-story hotel in Las Vegas.

What's worse than our faulty streak-detection capacity is another more subtle problem open to exploitation by a gambling game—our capacity to compute and respond to "what might have been." Almost all mobile creatures possess mechanisms that let them learn directly from their own experience and indirectly from information that tells them what they "might have experienced." It's this latter kind of learning, technically called counterfactual learning, that gambling games exploit with a vengeance, creating fictive error signals to which the decision machinery responds. We will revisit this same idea in the next chapter, where we show how market investments are influenced by covert counterfactual signals generated in your brain.

THE FEELINGS WE
REALLY TREASURE

Regret and Trust

Twenty years from now you will be more disappointed
by the things that you didn't do than by the ones you did do.
So throw off the bowlines. Sail away from the safe harbor.
MARK TWAIN

MARK TWAIN WAS talking about risk—or at least our tendency to overestimate it. We avoid risk because risk means loss, and as we have discussed throughout this book, our nervous systems are designed to avoid loss as much as possible. So we avoid risk generally, but not all risk. Risk is the companion of that other thing—trust.

You have to give someone the chance to cheat in order to learn to trust them.

So said U.S. Air Force officer, cognitive scientist, and moral philosopher Bill Casebeer. Once, he decided not to monitor his students during an upcoming exam. If there is no risk, there can be no trust, so Casebeer was arranging a situation where good information about trustworthiness could be extracted from the students' behavior. I never asked him exactly how he was eventually going to find out about the rate of cheating in the unproctored test. But his point is clear. Suppose that during a test I arrange a series of cameras that monitors each student, the students know this, and by implementing these measures, I remove all possibility of cheating. After the test, do I have more or less trust in the students' behavior? Neither. I am just as uncertain about how they will behave as I was before they took the test. Trust entails risking the loss of something valuable. In an exchange where you trust someone with your money, the risk has a calculable price. Trusting someone with your heartfelt affections, professing your feelings for her, also entails risk, but the calculation of this risk and the nature of the currency are not well understood. And it seems to me that there is also something ineffable influencing that kind of trust—something outside the reach of mere symbols. But for now the overall point is clear: Investing trust in someone else helps you model her. You gather information about her—and often extremely valuable information. Casebeer can only model the trustworthiness of his test takers if they have the option to cheat; that is, if he accepts that risk.

Our recognition of the truth in Casebeer's remark identifies only one of the many abstract emotions that we can all sense and that define us as social creatures: trust, regret, fairness, love, aesthetic satisfaction, and so on. These feelings help define each of us as an individual, and they help organize important social structures like schools, neighborhoods, governments, and social policies. We are so psychologically close to these emotions that it's difficult to see clearly that they possess patterns—regularities that can be captured in mathematical descriptions—and that they are "about something." My initial feeling, when presented with the problem of modeling them as computational products of my nervous system, was that the task was simply way out of my league. To transform emotions into mathematical regularities just seemed too big a problem. But if this were really impossible, then how would DNA build the device that generates and experiences emotions? If there were no mathematical regularities in emotions, then why should we be able to share and understand these important states in one another? If there are patterns and regularities to our emotions, then there must be a formal way to describe them.

Over the last decade, my opinion on this has changed, and I have become convinced that we can capture regularities underlying emotions in equations, and seek their neural underpinnings and evolutionary origins. I suspect that this effort may leave something out, but that's OK, that something will become clear. Furthermore, some parts of emotions can be cast as important computational signals necessary to guide our moment-to-moment survival as individuals and our longer-term survival as a species. Instead of taking on the entire problem of feelings and emotions, let's address the kinds of covert computational signals that contribute to them. This maneuver strips away the feeling part and many of the interesting complexities, but it leaves us with something tractable. Even the important, complex experience of trust can be stripped down to its bare essentials and viewed as a signal that updates our internal models of other humans, an essential capacity for any flexible, social species, and a capacity known to break in humans.

It's tempting to frame emotions as evolutionary adaptations to the learning problems that faced our ancestors. Reasonable scenarios can be conjured to support this contention, which is basically correct, but it misses something. I want to describe something a little more proximate, and generate mechanistic questions that pertain to our

ongoing behavior now. I'm not arguing that evolutionary explanations are not necessary. Nothing about computational descriptions of
emotions can really be understood without an understanding of their
evolutionary origins and functions; however, by keeping our focus on
the proximate we can learn more about how the brain computes emotions in a modern human, how these computations interact with our
complex culture, and ultimately how we make decisions now.

Let's return to goals. One central issue lying behind the idea of the
willful control of behavior is the concept of a goal. Merriam-Webster
defines a goal as "the end toward which effort is directed," that is, an
aim or purpose. However, the unadorned idea of a goal is ambiguous,
since goals exist at many levels and differ depending on one's point of
view. I have discussed how goals can be supported in our brains, can
guide behavior, and, in the case of the superpower, can act like primary rewards. *Proximate goals* are goals that impinge directly on immediate survival. *Ultimate goals* relate to the *fitness* of a species, where
fitness is the relative probability of survival. Ultimate goals are not restricted by the proximate needs of a particular individual, but include
other individuals in other generations and in other species. Ultimate
goals are something that the economist Adam Smith would recognize: They are the invisible hand of fitness, and they don't really care
about individuals.

The division of proximate and ultimate is an operational conve-
nience; however, it produces useful hypotheses about goals that inform the style and structure of formal models of emotions, especially
reinforcement learning models. One hypothesis is that proximate
goals are explicitly represented by behavioral algorithms and medi-
ated by identifiable neural systems (albeit distributed ones). This expectation arises because proximate goals are about immediate needs

of an individual; e.g., eating enough, drinking enough, avoiding danger, noticing potential mates, etc. Hence, behavioral experiments probe indirectly an individual's proximate goals, and physiological measurements that accompany such experiments concurrently probe the mediating neural substrates.

The influence of ultimate goals is different and more subtle. The reason for this expectation is the nature and scope of their influence: They influence the overall fitness of a species. We would expect these goals to be expressed in the particular style of behavioral algorithm or neural implementation that, when played out across generations, influences the overall fitness of the species. This indirect "stylistic" influence presents two important possibilities: (1) Ultimate goals may at times run counter to the immediate needs of any particular individual in a species, and (2) ultimate goals are probably represented implicitly, but not explicitly, by specific neural systems or algorithms.

Proximate goals also provide an ordering strategy for the control signals that influence their acquisition. As with any complex information-processing system, there are many levels of software in the brain. There is no more important proximate goal than to survive to see another day. All other subgoals, like "Get food if hungry," "Avoid painful things," and "Chase sex appropriately," take a backseat to this most important proximate edict—"Stay alive until tomorrow." For example, the goal of avoiding pain can be easily overridden if bearing the pain keeps the creature alive—we will hang on to a painful thorny branch if letting go means that we fall to our death. Here, neural commands that prevent release from the thorny branch veto control signals associated with avoiding pain. One goal supersedes the other. This prioritization shows that goals provide a natural way to order the level or precedence of their error signals. We know very little about how the brain makes hierarchical goals, but the best guesses place the hippocampus, striatum, and prefrontal cortex at the center of constructing and maintaining goal hierarchies.

BRAIN IMAGING EXPERIMENTS have now begun to probe the expression and repayment of trust across two interacting humans. One new twist in these experiments is the capacity to scan both brains at once. This hyperscanning of both brains allows researchers to go back and look at the correlation of neural signals across the two brains. This technology has already provided some support for the notion that ideas can gain reward status by acting in the brain like a reward signal.

Let's drop in on a modern version of one of these trust experiments. We enter a control room at Baylor College of Medicine in Houston, Texas, to face the belly of a superconducting magnet visible through a magnetically shielded window. A human subject is slowly being inserted into the machine as if being swallowed by the gaping mouth of a large sea creature. This is a functional magnetic resonance imaging device, or fMRI for short. fMRI is a descendant of magnetic resonance imaging (MRI), a noninvasive technique for imaging the inside of the human body. MRI takes "snapshots" of your knee, liver, or brain and produces those familiar grayscale images. Functional MRI uses the MRI device in a different mode of operation and instead of static snapshots, it makes "movies" of microscopic blood-flow changes in your brain. Microscopic blood-flow changes are tightly correlated with nearby changes in neural activity in the brain. By eavesdropping on the blood-flow changes, we get an indirect "movielike" measurement of changes in neural activity. This is a very safe procedure with no radiation or needles. A subject can be put in a scanner and asked to imagine his dog, remember a grocery list, or play a game that measures his risk profile with money. As with any new scientific technique, there are technical and interpretive issues hovering around all fMRI experiments, but for cognitive science and neuroscience, it has been a real godsend. Hyperscan fMRI (h-fMRI) moves fMRI naturally into the social domain. It's a software system that links fMRI scanners and other devices together over the Internet to allow simultaneous scanning of interacting brains. This kind of experiment has now been carried out across the continental United States and in similar experiments between the United States and China and Germany.

Just as the Houston subject's head disappears into the bore of the magnet, we notice a map of the United States on the computer screen with arching lines connecting Texas to southern California. After structural scans have been run on the subject (normal MRI), there is a pause and a voice with a mild French accent interrupts over an open voice link: "Ready to initiate autostart sequence." The voice is that of graduate student Cedric Anen at Caltech's neuroimaging center, where another subject has been inserted into a similar device. There is a pause, followed moments later by a high-pitched warbling sound. "Hyperscan started," sounds out back in the Houston control room, but this time delivered in the east-Texas drawl of graduate student Damon Tomlin, Cedric's cross-country counterpart. This is a hyperscan experiment, and the participants are half a continent apart in Houston and Pasadena, playing a game, while their brains are scanned. The two subjects trade money in a kind of economic exchange called a trust game, where trust and its repayment, trustworthiness, are literally measured in units of dollars. "I love you," "The check's in the mail," "I'll gladly pay you Tuesday for a hamburger today." . . . These are all versions of, simply, "Trust me," but trust is complex. Just consider the number of pages consumed by the average novel that deals with love, trust, and betrayal. *Trust* is hard to define broadly, but most of us can sense trust and deviations in trust just as readily as we can smell burning food. We are preequipped with a lot of neural machinery to detect and respond to trust in other humans and to use these sensibilities to guide our decisions. So broadly put, trust is a complex yet instinctual emotion. Trust is flexible and is subject to change based on the smallest piece of new information. And slight changes in trust are extremely informative when our brains try to model the intentions of others, so one of the functions of trust is learning, learning about others.

The word *trust* immediately brings to mind some relationship entanglement, yet trust is now being probed in behavioral and imaging experiments all over the globe. The expression and repayment of trust is a central social signaling mechanism that guides our behavior and is particularly important for our capacity to cooperate. Social psychologists and economists have long recognized the need for trust in social transactions, and it is well known in the mental health field that many mental illnesses cause dramatic changes in the capacity of afflicted individuals to trust others in a reasonable and adaptive fashion. Pathologies of interpersonal trust abound in our society and are also commonly observed in addiction, autism, schizophrenia, depression, bipolar disorder, and numerous personality disorders. Despite its widespread influence on human interactions, trust can be stripped down to its bare essentials, packaged into simple social interaction experiments with human subjects, and studied in a scientific manner. Of course, these experiments touch only a small sliver of trust's scope; however, the importance of understanding trust reaches beyond the laboratory and into the lives of families with members who cannot build proper social models of other humans, misattribute the intentions of others, and are generally unable to trust others.

One player in the fMRI, called "the investor," is given twenty dollars at the start of each round. He makes a choice—keep the money or invest any fraction of it into a "pot" that delivers a return of three times the investment. The second player, called "the trustee," chooses to divide the pot between himself and the investor in any way he chooses. Round over. The next round, the investor starts again with another twenty dollars. The players are told they will make an amount of money scaled to their performance, so there is real skin in the game. The players carry out ten rounds of this exchange.

On a round of play, the investor might give ten dollars, which grows to thirty dollars, and the trustee might choose to split this appreciated amount in half—fifteen dollars for each player. The investment phase displays trust by the investor and encodes it into money units. The repayment of this trust by the trustee, the trustworthiness, is encoded in the proportion of the appreciated amount returned. The give-and-take nature of this game allows each player to build a reputation in the mind of his partner. This experiment has actually been carried out using more than two hundred subjects (a very large fMRI study), and two remarkable findings emerged. First, reciprocity, exchanging good for good and bad for bad, was the behavioral signal that dominated the way the players responded to one another. A strong neural response for reciprocity showed up in the caudate nucleus, a region where numerous signals important for learning and behavioral control converge.

Second, a signal appeared in this same region (caudate) that correlated the intention of the player to either increase or decrease his level of trust *on the next round of play*. More remarkably, this "intention to increase trust" signal had all the characteristics of a reward-prediction error signal. One interpretation is that the idea about what an increase of trust would do to the partner's response was itself acting as a proxy for value. So an error signal in deviations from neutral trust, tit-for-tat behavior, showed up as a reward prediction error. This is exactly what one would expect if the idea of trusting the partner with more money came to act as a reward! This is the first hint from direct neural evidence that ideas like trust gain value in the minds of the trustor and trustee by acting like a reward, using the superpower trick, plugging an idea right into the reward socket. And the brain responses suggested that the brain went ahead and computed a prediction error in the expected trust, broadcast this to the rest of the brain, and used it to improve its predictions of future trust by the partner. Put another way, the brain used a prediction error signal in trust to develop a model of the partner—and we were able to watch this forming using hyperscan fMRI. Like any other finding, many details remain to be worked out, but it looks like h-fMRI is a feasible method to probe social exchange and it may have given us our first glimpse at a controlled use of the superpower.

As I was discussing these new trust experiments with my students and postdoctoral fellows, a great argument broke out about whether Chinese players would play the game differently. My sense was that they would not, but the Chinese students disagreed strongly. After a couple of weeks of Arguments-R-Us roundtable discussion, I finally decided to let nature answer the dispute. We would carry out the first intercontinental trust game with China and simply settle the issue behaviorally and neurally. There were several technical issues involved, including the time difference with China and possibilities for larger-than-normal Internet delays.

There was an extra layer of social interaction involved to pull off such an enterprise. We were fortunate enough to stumble upon the generosity of Paul Chu, the president of Hong Kong University of Science and Technology (HKUST), pioneer in superconductivity, and recipient of the National Medal of Science. A personal contact put me in touch with Paul; he visited my lab, the Human Neuroimaging Lab at Baylor College of Medicine, and I think he was pleasantly surprised by the tapestry of intellectual talents that we have been fortunate enough to attract in recent years. The group is far from standard at any medical school. The Human Neuroimaging Lab was built through a philanthropic gift to the school and through the farsighted vision of several members of the administration—Stuart Yudofsky, chairman of the department of psychiatry, Jim Patrick, dean of research, and Peter Traber, the college's new president. The stated focus of the Human Neuroimaging Lab (HNL) is social cognition; however, it's being pursued with a twist. The basic idea is that many forms of mental illness and many of the most important questions in psychiatry, neurology, neuroscience, and cognitive science intersect in human social cognition. The guiding perspective is that the biological part of the problem is different from the computational part of the problem, and it's at the connection between these two descriptions that real progress will be made. The group is a patchwork of expertise: mathematicians, physicists, psychiatrists, computational neuroscientists, economists, social psychologists, neuropsychologists, computer scientists, and engineers. The point to this diversity that social cognition is too hard a problem to attack solely from the bottom up or the top down, so we opted for both.

The Chinese trust experiments are a joint effort between the hyperscan development team along with several students and postdocs in my lab, and a collection of terrific collaborators in China. The experiments are actually being carried out as four versions of the same experiment. It's again a ten-round trust game as described above, except we are interested in how cultural influences may set "trust parameters" differently in Chinese and American brains in a way that we can see using hyperscan-fMRI. The first version of the experiment is being played anonymously between China and Houston. This experiment recruits Chinese volunteers with Web postings and posters written in Cantonese. The recruited subjects are instructed in Cantonese, and Chinese characters are used during the game to start new rounds and to solicit investment and repayment. Basically, we are indexing their "Chineseness" with these maneuvers. In a separate group, English is used for the recruitment and the game itself. Many answers are still out, but the group recruited and instructed in Chinese uses a different strategy, especially in the role of trustee, and we believe that there are a couple of different brain activations that correlate with this different style of play. The analysis of the experimental measurements will continue to emerge over the next several years, but I think that the effort shows the kinds of experimental probes of social cognition that will be possible in the future. It's pretty clear that I need to add a cultural anthropologist to the lab's intellectual fabric, since questions about how to analyze these kinds of data are not separate from an understanding of how cultural influences may act as brain-changing inputs.

————————

HYPERSCANNING TECHNOLOGY IS being distributed around the world in order to encourage others to enter the field. Social interactions form the backbone of every institution that we care about, family, business, school, government, and so on. Understanding how our brains contribute to the function or dysfunction of such units is important if we are to make wise choices about them in the future. Trust is just the beginning, as there are other powerful emotions that can be framed as computational signals important for learning and social behavior. The next signal is sadly familiar—regret. "You don't understand. I coulda had class. I coulda been a contender." These lines from the 1954 film, *On the Waterfront*, stumble straight from the heavy mouth of ex-boxer, present-day bum Terry Malloy, as played so convincingly by the then-young Marlon Brando. The comparison puts on clear display one of the most important learning signals in the human brain, regret. Malloy was comparing what he "coulda been" to what he actually was, a comparison of something that did not happen (becoming a real contender) to something that did happen (he ended up as a bum). We all recognize instinctively that for Malloy this difference is negative. Had the circumstances been reversed, the "sign" of the comparison would also reverse: "I coulda been a bum, but I ended up as a contender."

This classic morality play won all sorts of awards in its time, but for our purposes it illustrates what we all recognize as regret, a comparison of the possible (counterfactuals) with the actual (experience). At the psychological level, an enormous amount of work has been directed at understanding the emotion of regret and probing various forms of it: retrospective regret (regret about past events), prospective regret (regret over possible future events), and something that we might call near-present regret. Neal Roese wrote a terrific overview of academic work in the area that doubles as a self-help manual about how to transform "the road not taken" into real opportunities. I recommend it highly. However, the regret that I'm going to address is the less flamboyant cousin to the form of regret discussed by Roese. I'm going to focus on computational regret signals rather than the entire emotion of regret. Regret acts as the natural learning signal for counterfactuals, fictive learning signals, and can be used to unconsciously guide how we behave (our choices) and what we learn in important situations.

Let's return to Malloy for a moment and consider the following pseudoequation:

$$Terry\ Malloy's\ Regret =$$
$$(value\ of\ being\ a\ contender - value\ of\ being\ a\ bum)$$

The value of being a bum subtracted from the value of being the contender that Malloy could have become equals his regret. A more detailed mathematical version of this idea was proposed simultaneously in the early 1980s by Graham Loomes and Robert Sugden, and separately by Daniel Bell. In their formulation, regret theory is a way to model decision-making under uncertainty, that is, the kind of decision-making that we all have to do. It's my suspicion that regret theory was born as an attempt to produce a normative model of decision-making that could handle some of the oddities of real human choice, but mainly from the perspective of the emotion of regret. This theory has undergone significant development in the ensuing years.

For now, let's stipulate that there is a mathematical version of regret theory, and this formulation can be interpreted as casting regret in the form of a learning signal just like the ones we have been describing for experienced rewards, but with a twist. The twist is really two twists. The first is that regret must wait. Regret can be computed only *after an action* is taken and more information becomes available ("the road not taken") to change the value of "the road taken." Humans can form estimates of their anticipated future regret, but this is only an estimate based on what *might* happen in the future. The real learning signal *regret* is not available until after an action is taken. The second twist is that regret is typically inexpensive, and so it's alluring to any efficient computational device. Regret is good because it's informative and inexpensive; it signals a cheap way to get pertinent information from the other alternatives without experiencing them. No wonder our nervous systems use this information as a learning signal; it's almost like a free ride. The catch is that our brains must be able to frame the possibilities correctly. It would do no good to use regret over a set of possibilities that were extremely unlikely ever to occur for me in my lifetime. This means that regret requires good models of the world that I'm likely to encounter. This requirement suggests that regret signals in the brain will be intimately linked to gating signals controlling the way the prefrontal cortex frames problems and the possible choices that solve them.

Allie, a second-year medical student, was also a social "do-gooder," someone who really cared about those less fortunate and chose to act on those impulses. She was to become a pediatrician and her goal was to work with the children of the poor. But at the moment her goal was to make money for herself. A brief lapse in her moral universe, perhaps? Not at all. She was playing an investment game probing her brain's response to "almost wins" and "wins that might have been." The game was simple—watch a market for some time, invest some amount of money (or do the opposite), let the market play out for a short period to reveal whether a loss or gain had happened. Repeat. She surveyed the wiggling line that ran from the left-hand side of the chart to near the middle. Her hand tapped away at a button box resting on her chest, and we watched as her intended investment level rose ten, twenty, thirty, forty percent. She hesitated at forty percent, waited a moment longer, tapped back down to thirty percent, and submitted her bet. She was betting thirty percent of all the money she had available, which at the time was $185. She waited to see whether the market would rise or fall. She was tense and she did not like waiting for the next change in the market to reveal itself. The computer waited a bit and then displayed the market—it went up a lot; a gain for her. What do you think she felt? Was she buoyed by the gain or saddened by the extra gain that she could have made had she ventured more than thirty percent of her $185? Did her brain compute regret and did her emotional experience parallel that computation? This experiment probed the covert parts of the emotion of regret—the fictive learning signals in her brain.

Allie was playing a gambling game designed to probe brain responses related to fictive learning signals, that underlie the emotion of regret. This experiment, designed by mathematician Terry Lohrenz, was inviting. Subjects begin with $100 and if they play the game perfectly they can make over $500 within about thirty-five minutes. Terry, a mathematician by training and a working economist by inclination and experience, chose the biggest gambling game of all—the stock market.

Working with two other economists, Kevin McCabe and Colin

Camerer, we set up a simple investment game where subjects were endowed with $100 and "played" their investment decisions against twenty different markets. In the experiments, half the markets were "live," meaning that the subjects could make real investments and experience real losses or gains. The other half of the markets were experienced as "not live" and the subjects were asked to make visual discriminations about whether the market was going up or down. The decisions in both conditions required the subjects to make actions using button boxes resting on their chest. The trick here was the choice of markets. Initially, we thought that it would be fine to simulate the markets, that is, use some standard models that produce price changes that "look like" real price changes in real markets. Instead, we realized that it would be much more informative if we used historical market traces and measured brain and behavioral responses to these. This extra level of "realism" would let us ask an extra question.

What is the brain and behavioral response to known historical markets if we remove all the complicating cultural influences at the time the market was generated?

Our subjects would not know that they were playing their investment sensibilities against real markets, nor would they have any of the biasing influences of the times to navigate their choices. We were especially interested in markets with bubbles and crashes because we planned to carry out our next experiment using a model of market bubbles conceived by Economics Nobel Laureate Vernon Smith. The plan was to scan a group of people simultaneously while they created and responded to market bubbles that formed while they made investments. However, our first realistic market experiment provided some dramatic results on its own.

Two results jumped out of this experiment right away. First, some of the markets used were special. For example, the stock market crash of 1987 was particularly brutal in its capacity to cause losses for the fifty-two subjects who played the game in the scanners. But the mother of all markets was the epic crash of 1929, the one that initiated the Great Depression. None of the subjects emerged from this market with more money than when they began, and many of them lost more than fifty percent of their total portfolio. In fact, a professional economist who played the game recently lost this amount in the last choice on the last market; unfortunately, the last market was the 1929 crash. This market, out of all twenty used, lulled subjects' decision mechanisms into a kind of stupor and then—bang. Goodbye, money. The second hard-to-miss result was the variable that drove much of the behavior in this investment game on all the markets—regret. Not only was regret correlated with the way that subjects changed their bets, it also showed up as an extremely strong neural signal in a reward-decision-making region of the brain, the ventral putamen, the same site where reward-prediction error signals appear. So experienced reward error signals, which register the difference between expected reward and experienced reward, activate the same brain regions as counterfactual error signals, the difference between what could have been obtained and what was obtained. This confluence of experienced reward error signals and counterfactual error signals in the same brain structure is suggestive. Maybe this is why our brain treats counterfactual experience just as it does real experience: Both are sent for processing to a structure that may not differentiate between the two. The fMRI technology is far too crude at this point in time to make such a distinction, but it's a reasonable guess at this stage.

Why are these results significant? Let's remember what we concluded in the last chapter: Gambling games evolved to exploit the frailties of our biological valuation and decision-making machinery. They are crafted to "fit" the way our nervous system works, and they lull our decision mechanisms into a kind of addictive complacency. Normally, markets are thought to be influenced primarily by the mood of the time and the important financial or other events going on. However, this could not have been an effect on our participants. They did not know that they were playing historical markets, so there was no general mood of the moment to influence them. For the average investor in our experiment, the 1929 market was not a history-making event but just a gambling game against which they made a bet in an experimental setting. There was no cultural context, no zeit-geist, and no pressures of losing their life's savings. Yet every one of our subjects lost his shirt to the 1929 market! It was special even in this completely unrealistic lab setting. Somehow, the 1929 market hit a kind of fragile "sweet spot" of the valuation and decision machinery in the subjects' brains. And in Lohrenz's analysis, regret, the counterfactual learning signal, was a powerful determinant in the betting decisions that created their losses while playing against the 1929 market. It is not too bold to suggest that the investors' brains were also so influenced seventy-seven years ago during the real market. Other factors were surely also involved, but it's wonderfully surprising that the market's movements were already poised to fool the average investor's brain. Add to that any other panic-inducing, stress-inducing events of the time and you have a market pushed around by some labile decision-makers. The brain's machinery for sensing and responding to counterfactual data clearly seems to have contributed to the initiation of the Great Depression. And of course the full-blown emotion of regret was certainly experienced in 1929 during and after the crash. In fact, as cultural and financial historians run that period through various analyses, regret about a host of decisions, individual and institutional, permeates every one.

Lohrenz's experiment suggests some further practical possibilities in both the medical world and the economic world. The economic possibility is straightforward: Could markets be actively monitored using a stylized computational model of individual human regret responses, like a human regret monitor constantly probing how market movements are likely to influence individual human brains? The idea is to take a group of human subjects (average or expert) and measure their regret responses (behavioral and brain responses) by letting them bet on markets in the same way that Lohrenz did in his experiment. From these data, one could extract a human regret model that might be used to better understand how markets get out of control. It's not too long a reach to imagine this scenario in the near future. The current situation in the stock market is basically an uncontrolled gambling experiment with lots of players injecting diverse levels of influence on market movements—inside players, outside players, marginalized players, big players, and small players. I don't know anything about the stock market, so I can't offer anything resembling financial advice. My proposal is focused on the response of average human brains to market movements and changes in behavior that correlate with these responses. With current technology, it is reasonable to try to understand better the mechanisms employed by individual decision-makers that cause bubbles to form or prevent them from forming. Minimally, we should be impressed that Lohrenz's experiment has provided some new insights into a dramatic and historic economic event, one that we may want to prevent in the future.

These same experiments also suggest important new directions, and perhaps even diagnostic tests in the medical arena. A gambling game that generates measurable regret signals might indeed prove useful in understanding gambling addictions or even drug addictions. More importantly, regret guides lots of other important decisions that we make. It just happens to be easy to probe with a gambling game. The particular use of gambling is suggested by recent observations with the treatment of Parkinson's disease. Patients with mild PD were given drugs that mimic the influence of dopamine in the brain, in particular newer drugs that activate selective dopamine receptors. Dopamine receptors are the membrane-bound molecules that "read" the messages carried by dopamine. The drugs, called dopamine agonists, pushed a significant number of patients into a risk-seeking state. Many of them developed gambling addictions even though they had never before gambled! So there are very practical reasons to understand both the biological and computational underpinnings of neurological and psychiatric disease. I suspect that a subfield of computational psychiatry is just around the corner. It's also reasonable to ask if regret signals in the brain are perturbed by mood disorders where "what could have been" is mischaracterized (everything could have always been much better) or even by obsessive thought disorders where "what could have been" is the obsessive end itself—that is, pursued like a reward. Whatever the future brings along these lines, these possibilities show that theoretical models of counterfactual learning signals provide a new perspective on normal and pathological brain function.

I HAVE MAPPED THE IDEAS of trust and regret onto quantifiable computational signals important for learning about the world and other humans efficiently. What about our internal sense of fairness? We have all experienced the feeling that something "just seems fair" or not. Sometimes we can connect the feeling to something taught or experienced in our youth, but usually the feeling is just there, impelling us away from one choice and toward another. Is it also possible that these instincts can be captured by computational descriptions in the same way as trust and regret, and related to brain responses? It's entirely possible, so let's explore some of the work by behavioral economists and neuroscientists uncovering the proximate mechanisms behind these flexible instincts.

One approach to fairness comes from the world of behavioral economists who have used simple economic games to probe our sense of fairness. One such game is called the ultimatum game. It's played by two people and is a one-round bargaining game that is better called "take it or leave it." An amount of money, say a hundred dollars, is made available to a proposer who offers some split of it with the other player (the responder). The responder can accept the offer, in which case the two players each take their share; or the responder can reject the offer, in which case both players leave empty-handed. Hence "take it or leave it." Rationally, all nonzero offers of the proposer should be accepted by the responder, since both players start the game with no money. Too bad for rationality, since players routinely reject nonzero offers and our instinctual reaction is that the rejection occurred because the offer seemed "unfair." But exactly what does this mean? Before the exchange neither player had any surplus money. Is it not rational for the responder to take any offer where she receives some amount of money? It's only rational if we ignore the fact that humans carry around fairly good models of one another. In a two-party exchange, we have a good model of what the other person is likely to do and we share this modeling capacity with our partner. Also, we share norms with our partner as well; that is, we both expect certain behaviors from one another and these expectations are what drives the proposer's split and the responder's willingness to reject nonzero offers that are too low.

This kind of experiment has been carried out across many different cultures and for different levels of stakes. The results are basically the same—positive offers are rejected and the rate of rejection increases as the offer size decreases. Surely we are seeing a fairness instinct at work. Well, there is at least a fairness instinct at work, but I suspect that there is something a little craftier taking place, but just out of the reach of this one-round exchange. The craftiness is a computation that the brain is probably doing—it's not just computing its own outcome but also that of the partner, and making projections about both into the future. The bad news is that a one-round exchange will not reveal such a subtle calculation, and that's part of the limitation with the ultimatum game. The good news about it is that is has provided some of the first neuroimaging data on the human fairness instinct.

A number of prominent experimental economists have carried out these games and one feature that emerges time and again is that humans will indeed punish "unfair" offers at a cost to themselves—the "leave it" option where everyone walks away with nothing. But from the brain perspective, it's the pairing of this game with neuroimaging that has opened new doors. Jon Cohen's group at Princeton University was the first to pair functional imaging of the human brain with the ultimatum game and show directly how brain responses correlated with the degree of fairness of the offers made by the proposer. Although multiple brain regions were active when a subject rejected an offer, one particular region was particularly predictive of rejections—the anterior insula. Stronger activations in the anterior insula predict that the subject is more likely to reject the offer. This "rejection response" was interpreted by these authors as an example of a negative emotional response. Had this area been monitored during the experiment, one could truly have eavesdropped on the subjects' intention to defect—at least statistically. It would have been impossible to say with certainty whether or not a rejection was imminent, but this kind of statistical "mind-reading" is certainly possible and has been carried out in a number of other studies. The anterior insula will also activate strongly when a subject has a "disgust" reaction to facial expressions of disgust (a sympathetic response) or pictures of disgusting things like mutilations or contaminations of various sorts. Collectively, these results suggest that something like the *metaphor of disgust,* or more precisely the emotional category of disgust, employs common neural machinery across the many dimensions in which disgust can be experienced. This metaphor-storage idea is a provocative one, since it represents a very powerful way to organize information across diverse dimensions. These are early days for such conclusions to be considered firm, but the possibility is real and the idea is being pursued.

As I noted, one major conceptual issue that surrounds one-round fairness games is that they are one-shot, one-round interactions. This isn't how humans used to interact. Our instincts for sensing and responding to fair exchanges evolved in a social environment where tit for tat was king. What you did to me today was coming back to you tomorrow in kind. Social exchanges of one sort or another were only rarely one-round interactions; instead, our forebears cooperated and competed through iterated exchanges. In modern society, one-shot interactions are common, and long iterated exchanges are less so. Although we may see our neighbors frequently, we may never experience substantive give-and-take exchanges, a fact sadly true in many American suburbs. However, I doubt the recent increase in social isolation has influenced the basic components of our social exchange algorithms. It's my hunch that one-round exchanges unnaturally induce odd behaviors in the average human brain and that much of the strangeness in ultimatum-game responses is due to different depths of search in the minds of participants—"I know that she knows that I know that . . ." Before we adopt the one-shot exchange as the gold standard for reciprocal interactions among human beings, we will need more empirical proof of its validity. I suspect that the coming years will see definitive behavioral and imaging experiments, because there is a tremendous groundswell of interest in this field and cooperation is such a fundamental human behavior.

What about specific reactions to fairness or unfairness? How willing are we to act to impose a cost on unfair behavior? In the ultimatum game, we saw that the rejection rate ("leave it" rate) measured those offers considered unfair, but as I noted, this may be asking our fairness computations to operate outside their normal range. However, it's pretty clear that humans feel the need to place penalties on unfairness. This idea has long been embedded in our legal system;

treating someone unfairly in an economic exchange is punishable by law. If you take food from the grocery store without paying, the law recognizes this fact with a penalty or imprisonment or both, depending on the severity of the theft. The law also frowns on the unfair act of taking money from someone through deception, and encodes this societal scowl as fines and terms of imprisonment. The importance of understanding the origins of our fairness instincts derives from the many ways that fairness is used to define the quality of our acts, laws, rules, and social organizations in general.

Our collective social contract recognizes the degree of reciprocity as well as the overall value of it. When you take food from a grocery store, you exchange money for the food. You are arrested for shoplifting if you take the food without paying, but you are also arrested if you take the food without paying enough, usually the listed price. And although local jurisdictions will differ in the exact scale used, the degree of underpayment is an important factor in considering the scaling of the punishment. So here we have an important social evaluation, encoded as monetary differences, and enforced by monetary penalties or temporary loss of freedom. I'm no legal scholar, but these facts make it paramount for us to understand how individual human brains measure such fairness scales and the degree to which individual human brains may differ naturally in their capacity to sense and respond to fairness. In a remarkable confluence of name and interest, some of the most important recent work on fairness has come from the creative work of Ernst Fehr (pronounced "fair").

Ernst Fehr is an experimental economist interested in, among a range of subjects, fairness and the behavioral mechanisms that humans use to enforce it. Experimental economics is a field that uses controlled laboratory experiments to gain insights into theoretical economic concepts and to better understand the economic decisions made by humans. As a field, experimental economics illustrates the highly integrative nature of scientific work on human decision-making and the growing role played by our understanding of brain function. Experimental economics was pioneered initially by Reinhard Selten, Vernon Smith, and Charles Plott. Reinhard, talented at both experiment and theory, shared the Nobel Prize in Economics with John Nash, known in some circles for his proof of the embedding theorem and to moviegoing audiences for the depiction of his struggles with mental illness in the film *A Beautiful Mind*. Smith has also won the Nobel Prize in Economics, but expressly for his work in experimental economics. Smith is a strong proponent of modern efforts to relate economic decision-making to neuroscience, a new field now known as neuroeconomics. To complete the web of associations, Smith's prize was shared with the multitalented Daniel Kahneman, who was awarded the Nobel Prize in Economics for his work in behavioral finance and behavioral economics. Kahneman is one of those special intellects whose interests range broadly and who make singular contributions to many fields. He enjoys respect in many areas of psychology, cognitive science, economics, and neuroscience.

Let's return to Fehr and ask how he frames his problem—the problem of fairness. Fehr has used a series of economic-exchange games to probe the human instinct to be fair and to punish those who aren't. The twist is that Fehr has produced experimental results that suggest that humans punish defectors at a cost to themselves with *no apparent gain for themselves*—his claim is that humans and humans alone possess an instinct to be altruistic. Not reciprocally altruistic, this for that, but good to others at a cost to themselves. The father of the modern idea of reciprocal altruism, Robert Trivers, remains skeptical of this interpretation. Consequently, this is a bold claim and Fehr has marshaled a series of experimental results to support his case. This work takes numerous forms, including ultimatum games, ultimatum games with competition, and something called a public goods game. The theme that surfaces in all his results is that humans appear to have a need to cooperate and will do so at a cost with no direct gain to themselves. For an evolutionary biologist, this claim on its face seems as if it has to be wrong. For the story developed in this book, that proximate brain mechanisms evolved under the stiff need to be economically efficient, the claims are equally hard to swallow. Fehr's experimental findings are extremely convincing; it's their theoretical interpretation that is in dispute—the normal state of affairs in science.

The simplest countering idea is that altruism is really self-interest in disguise. Indeed, this possibility can be exposed by considering the ultimatum game, not as a competitive game between two players, but as a learning problem where two brains learn at once. Let's return to the single-shot ultimatum game to understand what might be going on. One quantitative description offered to explain the ultimatum game results is inequality aversion, a built-in dislike of inequitable splits between the two players. Suppose I play the reader in an ultimatum game where I'm given $100. The idea of inequality aversion is that I dislike a $95/$5 split favoring me, but I also dislike a $5/$95 favoring you, although I dislike the split favoring you a little more. Inequality aversion captures what subjects do, but it sidesteps the important question about fairness—what's it about? Why should players have a fairness instinct that induces them to sense unfair splits and act on that sensation, even unfair splits that favor their own payoff? This is odd unless the subjects generate signals related not to just one round of play, but multiple rounds of play. The claim I'm making is that players' brains are unable to look at only one round, but instinctively make calculations over multiple rounds that "might occur" in the future.

Let's take the case of the $5/$95 split favoring you. The inequality aversion description says that the value of my payoff ($5) is reduced by the difference between your payoff ($95) and my payoff ($5) times some constant. To be concrete, let that constant be 1/20. If your payoff is bigger than mine, which it is here, the formula for inequality aversion reduces to:

$$\text{Value of my payoff} = (\text{my payoff}) - 1/20 * (\text{your payoff} - \text{my payoff})$$
$$= (\$5) - 1/20(\$95 - \$5) = \$0.50$$

The claim is that my brain reduces the value it assigns to my "winnings" depending on the inequality of the split across the two players. Here, my take-home pay is actually $5, but my inequality-averse nervous system doesn't let me enjoy that—it makes me feel like I received $0.50. Notice that if instead my brain really valued the inequality, it might set the constant over 1 instead of the 1/20 that we used here—in that case, I could actually win but feel like a martyr whose perceived payoff is next to nothing or even negative. Each win would punish me. If the constant were negative, I might wrongly be considered a saint, reveling in my losses because of their perceived value to me.

I have told only a reduced version of the story. The full-blown model for inequality aversion reduces my perceived payoff by the inequality felt by me *and* you.

$$Value\ of\ my\ payoff =$$
$$my\ payoff - A(your\ payoff - my\ payoff) - B(my\ payoff - your\ payoff)$$

Two important issues arise. First, the reduction in the perceived value of my payoff is best considered from a learning perspective, but a counterfactual learning perspective. The differences between your payoff and my payoff will induce regret signals, that is, counterfactual learning signals. The first difference is my regret, the difference between what you got ("what I could have gotten") and what I actually received. The second difference is your regret, the difference between what you "could have gotten" and what you actually received. So we could rewrite this as:

$$Value\ of\ my\ payoff = my\ payoff - A(my\ regret) - B(your\ regret)$$

Why should my brain consider your payoff as something "that I could have had"? Simple, because you are like me and you actually received the payoff. What's possible for you should be possible for me, so my brain perceives your gain as my possible gain. It's simply a way to learn a better model of the world, and here the world includes the payback from a partner in a kind of fake trade. The situation is directly analogous to the bee example earlier. Each bee uses the experience of the other bee to generate a regret signal, a difference between "what could have been" and "what was actually experienced," to build a better model of the nectar yields from the green and red bushes.

This is efficient, since each bee gets the informational benefits of two flights for the price of one actual flight that it makes.

Here, the yields depend on the other person, but they also depend on the person's brain treating the exchange as though it might extend for more than one round. In that case, the problem for each brain is clear—reduce the regret in both brains. If I choose so that your regret is high, then your perceived payoff will be lowered and you will be less willing to give more (or accept less) on the next round. If you choose similarly, then my regret will be high and I will become stingy. The solution, especially since both brains are trading exclusively with one another, is to choose so as to minimize regret across both brains for as many "rounds" of trade as each player expects. Optimally, we would like the joint regret into the near-term future to be zero, since that brings the highest perceived value to each player. This is of course impossible, so some compromise is reached.

Let's return again to our bee example. If each bee finds a way to reduce its regret, then they have all learned a better model of the field. They can anticipate the yields from flowers that they choose to experience *and* the flowers that they might have experienced. And the regret signals near zero are exactly the indication that the "actually received" plus "could have received" model is accurate. This is my claim for what the humans are doing on the ultimatum game and why this odd one-round game yields inequality aversion. However, understanding the inequality signals as counterfactual learning signals is an important difference, since it gives a reason (build better models) for its existence and it yields a way to understand multiround trades for all sorts of commodities. So the work on fairness, by Fehr, has led to debate and model-building across the neuroscience and economics divide and may lead further to a real alliance between economists and neuroscientists. I think that the resulting "fusion field," neuroeconomics, has quite a future and likely in ways that can't clearly be foreseen.

I HAVE SINGLED OUT regret and trust in this chapter because, as I have indicated, there is now good evidence that one can monitor brain responses correlated with the computational signals that contribute to these emotions. This capacity should help us start to differentiate the way that the brain implements these emotions. But there is another reason that I have not yet mentioned. Trust, and the ability to sense and respond to counterfactuals like regret, were essential capacities for early cultures to spread innovation within themselves or between two different cultures. Now, I will not here offer some facile theory of cultural transmission packaged in a few brief paragraphs; instead I am focusing on a few central *brain* capacities that would need to be working properly for such transmission to take place. I am once again focusing on the proximate mechanisms and sidestepping the ultimate influences that selected for the mechanisms. With that disclaimer in place, let's consider briefly how important interpersonal trust and counterfactual thinking were to innovation and perhaps to cultural progress.

In his book *Guns, Germs, and Steel,* Jared Diamond provides an invigorating account of why cultures advance at vastly different rates. In this work, he quickly dismisses silly theories of racial and intellectual superiority and makes his case for two prime movers behind differential advancement: the geography of resources and the transfer of innovation among groups that bumped into one another. On balance, geography is the grand-master constraint in Diamond's account, since it delimited migration patterns, and by "telling people where they could go," it forced patterns of climatic change with which the migrating bands of humans had to deal. There's nothing more challenging than having to handle major climate changes while still finding a way to eat enough. Diamond's arguments have been subjected to both reasonable and unreasoned criticism, but they provide a nice mechanism to highlight the importance of understanding the neural basis of trust, regret, and a collection of other social instincts (learning mechanisms). My purpose here is "brain-and-mind centric." I will focus on several mental features required of individuals that would permit the transfer of innovations. They fall into three broad categories: (1) the ability to notice long-term patterns, (2) the ability to simulate "what could be the case," that is, counterfactual thinking, and (3) the capacity to share and steal innovations.

Let's go through these three categories using the example of agriculture. Two epic technological steps for humans were the domestication of plants (the invention of agriculture) and the domestication of animals, facts highlighted by Diamond and analyzed by others. Diamond goes to great lengths in his book to emphasize the difficulty of domesticating plants and animals, and he takes care to catalog the rather limited list of species that could even be considered domestication candidates. For our purposes, let's turn this issue around and ask instead: What kind of mental capacities would permit such domestication? We will see that trust, regret (counterfactual thinking), and fairness play a large role in the answer.

First, there must be a capacity to sense long-term patterns in the environment, remember them, categorize them in novel ways, and value them. Cognitive science and psychology have long occupied themselves with the first three problems—the capacities necessary for extracting memorable patterns from direct experience. Accordingly, from the theoretical computer science side, those interested in machine learning have traditionally focused on those three. The last feature, the ongoing valuation and revaluation of memorized patterns, is also essential for cognitive innovation but has received vigorous work only over the last decade or so. However, noticing, remembering, and valuing patterns were indeed essential for the invention of agriculture. The first domesticated plants were annuals, plants that germinate, flower, and die in one year, and most domesticated grains like wheat are annuals. The seeds of wild wheat, when mature, fall to the ground, thereby planting themselves and completing a natural cycle. Domesticated wheat stays on the shaft when it's mature, a feature that allows it to be harvested efficiently. Now, it's argued that the "stay on the shaft" feature was a mutation that contributed to wheat's use in agriculture, but it's just as reasonable to suppose that there was phenotypic variation in the degree to which mature grains hung around on the shaft. What kind of creature could notice and remember the difference between the wheat that drops seeds to the ground and the wheat that keeps seeds on the stem? Furthermore, this creature would also need to remember the features of the wheat that advertised each property in order to be able to identify them again. Noticing and learning about these patterns required some seriously sophisticated working memory and the capacity for abstract thought. But even these features on their own would not be enough. To actually innovate, this same creature must be able to simulate series of what-ifs; that is, something like "What if I were to collect up the stay-on-shaft wheat seeds and spread them out nearby?"

The capacity to simulate a complex series of counterfactual what-if steps was necessary for early humans to even think about possible innovations. This should not be too surprising. We saw earlier that even single-cell bacteria can model their external environment and maintain short-term memory of how well they have been doing collecting food and avoiding noxious substances. I think that one key difference in this case is that the counterfactual thinking required simulation from both a first- and third-person perspective, something probably impossible for a bacterium. "What if I were to spread the seeds nearby?"—first-person perspective. "When he spread seeds nearby a lot more stay-on-shaft wheat resulted; maybe that would work for me"—third- and first-person perspective. Obviously I'm making up the specifics here, but the capacity to simulate actions and events not experienced, value the outcomes of these simulated possibilities, and use both to choose actions that "I" could take and "you" could take was essential to organize agriculture. In the market investment game previously detailed, we saw that the difference between "what could have been" and "what was" (regret in this case) activates brain structures in much the same way that errors between expected experience and actual experience do. So not surprisingly, a natural learning signal for counterfactual learning can be seen in imaging experiments in humans. This is only the first of such signals detected and the future should bring a host of others, providing connections between neuroscience and the social sciences.

The third requisite mental feature is the presence of a receptive audience, that is, a group that shares enough similar mental machinery to be able to capitalize on the cognitive innovations thought up by others. The capacity to share or exchange innovations is trade, but trade is potentially risky. The individual with whom you trade may take all you have and run. He may hurt you in the process. Trade requires trust and there is no risk-free way to trust another individual. So there must have been a way for the early hominid brain to value trusting other humans. We discussed one mechanism by which trust comes to have value—let trust act like a reward. But of course there must be calculated limitations to this process, since complete trust is too easy to exploit. The rudiments of the instinct of sharing are seen throughout social species. Vampire bats share blood, chimps share meat, and humans share nearly everything, but especially information. And the sharing is essential for cognitive innovations to be transferred throughout a cooperative group. Stealing is another way to share innovations; however, the theft of something tangible—meat for instance—is rather easy to detect and can be costly to the thief. Taking something like a trick for making a fishing spear or a new weapon is hard to detect initially, since the information may be taken by just watching or listening.

Economic exchange, broadly conceived, had to rely on the value of trust in the nervous systems of those participating in the exchange, a value conferred by parts of our nervous system that we are now beginning to map out. Not only has Jared Diamond emphasized the value of trade in the spread of technology, other authors have made the case that the capacity for economic exchange may have been the feature that separated the early modern humans from Neanderthals. Such a capacity requires trust, and trust is now subject to direct physical probes in modern humans.

FROM PEPSI TO TERRORISM

How Neurons Generate Preference

We live in a sea of cultural messages,
some more compelling than others.
ANONYMOUS

INDEED, WE DO LIVE IN a sea of messages. "Buy me," "Look like her," "Has been shown to cause cancer in lab animals," "Mind your mother," "Make more money"—and the mother of all messages, "That may go on your permanent record." Some messages we ignore and others imprison us, locking our minds on a perturbed view of who we should be. Cultural messages take many forms and motivate lots of different behaviors, but one thing is certain: They feed information and innovation back into our brains. Sometimes the "core" of a message is implicit:

A salad of perfectly grilled woodsy-flavored
calamari paired with subtly bitter pale green leaves
of
curly endive and succulent petals of tomato flesh in
a deep, rich balsamic dressing. Delicate slices of

pan-roasted duck breast saturated with an
assertive, tart-sweet tamarind-infused marinade.

Hungry? Two researchers wrote this wonderful passage in the journal *Neuron*. They were profiling neuroscience experiments on pleasantness processing. Unless you are a vegetarian, and maybe even if you are, the lushness of this description ignites all sorts of memories and brings to mind appealing foods and experiences. This entire message acts as a proxy for something of value, a promissory note of what is to come. Perhaps this passage drives you to salivate and make quick plans for the nearest fancy restaurant or perhaps it only slightly increases your thoughts of food. In either case, the message has done its job—to promise some future rewarding experience and make you more likely to want food.

Who hasn't bought a car or jacket or personal electronics based at least in part on a catchy advertisement that boils down to a singular message: "This is cool," "This makes you look skinny," or "This can hold a billion songs"? How do cultural messages direct our purchases? Cultural messages, like any other object processed by our nervous system, may come to direct our behavior in exactly those ways we have discussed. They gain value the same way that the "light predicts juice" scenario passed value from the juice to the light. Independent of any specific brain, cultural messages also have another agency that stores them, communicates them, modifies them, and associates them with other messages—culture. Cultures are storehouses of cognitive innovations and as such also represent forms of information processing. Although generated by the collective actions of lots of brains, cultures have storage and processing capabilities not possessed by a single human. They remember more than any single human, and

this collective memory feeds back dramatically to influence individual behavior. The social sciences have been studying and analyzing
cultures along these lines for many years, recording and interpreting
their products. This interaction between brains and culture is complex, and cognitive science and neuroscience are a long way from
making dramatic inroads, but experimental probes using cultural
messages have begun to emerge.

I would like to address a historic cultural moment in my generation's recent past—the Pepsi Challenge. It was not as grand a battle as,
say, Agincourt or World War II; it was more of a cultural skirmish between two corporations, a street fight, you might say. Participants in
the Pepsi Challenge were culled from shopping centers and malls
across America. Beamed to our living room televisions, a soda drama
was enacted before our eyes. A subject would be placed in front of
two unlabeled cups, one containing Coca-Cola and the other Pepsi-
Cola. The subject would sip from each cup and choose one, presumably the one that she "liked" best. At that point, a moderator would
cheerily reveal the contents of the cups, and, typically—as I recall—
Pepsi would win. Surely, Pepsi had a better flavor, right? And what exactly was the point of airing these commercials? I can only guess that
someone wanted to prove that "all colas were not created equal" and
that "content matters." Conclusion: In the content department, Pepsi
must be superior. But something was left out of the test that is normally present: the brands.

Brands really count, especially when our choice selects one to the
exclusion of another. No one puts his money into a soft drink machine and happily selects the unmarked aluminum can or plastic
bottle—and for good reason. It could be an enormous waste of money,
since no one knows what, if anything, the unmarked can contains.

Brands count because brains have been "branding" reward-predicting experience ever since there were brains. And in the case of soft drinks and foods, the brands predict future rewarding experiences (consumption of the contents). This is exactly the idea behind the reward prediction systems discussed in previous chapters. But here's the twist: A cue (like a brand) that promises future reward also influences the experience of the reward. In the case of caffeinated sodas, this claim is not particularly surprising. Brands act like cues that predict reward (soda contents); they generate expectations about the experience the contents will deliver, and that experience depends in part on consuming caffeine. Depending on the familiarity and value of the brand, these "brand cues" change dopamine delivery to various brain regions through their effect on reward prediction circuitry. But the consumed caffeine, following right on the heels of the brand-cue experience, changes dopamine levels on its own. Caffeine clogs the molecular pumps that vacuum up dopamine and therefore slows the ability of neural tissue to clear away this powerful neurotransmitter. These two effects on dopamine add up, one due to brand-induced expectations and the other due to caffeine's inhibition of the molecular maid service that cleans up the dopamine. A familiar brand of caffeinated soda should release more dopamine than if the brand was unknown and the drink decaffeinated. Consequently, we should not be surprised that these different physical pictures cause a change in the perceptual experience. On general grounds, we should not be surprised that familiar brands taste like something! Sounds remarkable on the surface, but these claims would not surprise an expert in flavor perception. They know that flavor is a composite perception built by our brains. It's built out of taste (sweet, sour, bitter, etc.), texture, temperature, delivery (think "presentation" in a restaurant), and expectations dug up from memory. Flavor, like every perception, is a complex mixture of information, including brand information, and like other perceptions, it is subject to illusions.

EXCEPT THAT THERE ARE no illusions, only normal operating ranges for different perceptions. One perception is essentially the same as another. Your brain is constructing everything, from your perception of time to whether or not your keys "feel" heavy in your pocket. And these constructions depend on many sources of information—never just one. Illusions occur when the minds that run on our brains hit their specified operating limits—the ranges of inputs and outputs over which they were designed to work efficiently. Press the system a little bit—ask it to perform at or beyond its limits—and there is what we call an illusion. Lines look longer, objects look bigger, leftward motion looks like rightward motion, front and back get confused.

A perspective drawing gives the illusion of three dimensions on a two-dimensional plane, but there is no third dimension present—your brain is combining information about the angles and continuity of the lines to construct the most likely scenario "out there." Until artists discovered tricks about how to emulate depth in two-dimensional drawings, humans did not realize that depth was not simply "out there," but was the mind's best guess at the likely "out there" given the current input. Scientists, especially visual system scientists, have exploited illusions for many years to probe the way important perceptions are constructed. Illusions aren't limited to vision or even only to our five senses, although they were discovered first in these sensory domains. We can experience time illusions where an effect is perceived to occur before a cause, demonstrating a remarkable flexibility in our mental software that has only recently been identified. We can also have illusions in the social domain. We can perceive someone as being more trustworthy than his actual behavior supports. While in love, the mind can construct a person that almost doesn't exist—friends standing on the sidelines routinely gape in amazement that we don't "see" the bad girlfriend or boyfriend behind the romantic vision.

This problem even plagues physical theories—they must be put in a form that can be consumed by our nervous system and the cognition it conjures, or at least the nervous systems of well-trained physicists with specialized knowledge structures. Nevertheless, limitations in our perceptual apparatus make difficult the physicist's job of separating what's really "out there" and what is being constrained by human psychology—the concept of time is difficult for just this reason. To improve their reports of what's really "out there," physicists could include better formalized models of our psychology so it can be properly subtracted from the report.

So AWARENESS OF A food's brand directly affects our perception of its flavor. How much can brand information and the internal "brand image" change flavor preference and hence our choices? Let's return to the Pepsi Challenge mentioned earlier. The "idea" of Coke or Pepsi has measurable influences on both brain and behavior. It clearly influences what people buy, and since these choices are not mysteriously generated, there must also be some underlying pattern of neural activity supporting the choice. These general conclusions are straightforward. The added twist in recent years is that we are now equipped to explain some parts of brand value in terms of neural systems—neural systems that are attacked by disease and exploited by drugs of abuse. This is the value of making a connection to the neural underpinnings; we can now ask directly whether cultural messages like brand images engage reward-processing and decision systems in a pathological way—for example, remove or diminish our capacity to choose. This is important both neurally and culturally.

It's particularly important to understand the neural and behavioral influence of advertisements and games pitched at children. Are these capable of damaging normal reward-processing and choice mechanisms? Do they exert undue behavioral or neural influence at particularly critical times in development? Are the effects long-lasting?

Brands, especially those associated with food and drink, should engage reward harvesting mechanisms in our midbrain, striatum, and prefrontal cortex. Just like the "light predicts juice" scenarios from before, a brand is a predictor that accrues value provided that it consistently predicts a future rewarding experience. And a high-value brand can pass some of its value on to other symbols with which it is consistently associated. A brand can also be devalued or even have its value completely removed in an instant (think Enron), again using the same general neural machinery. Choose a brand that subsequently makes you sick or causes a bad experience, and, poof!—good-bye brand value—especially if it makes you sick. This is characteristic of an efficient learning strategy—your brain uses the symbol (brand) to represent future sickness and so avoids investing in something that will not reward you but may potentially put your survival at risk.

One of the first imaging studies in humans to probe the neural impact of brands came out of Germany with the work of Manfred Spitzer and his colleagues. They asked whether a brand category, like sports cars, engages the reward-predicting and -harvesting machinery discussed in Chapters 4 and 5. Their underlying hypothesis was that sports cars have come to represent symbols of value in Western culture through their relationship to social status. Spitzer's group measured the influence of the "sports car cue" on the reward prediction circuitry in human brains. Not surprisingly, they recruited the most likely candidate brains for seeing such responses—young, healthy males. The basic idea was to contrast the brain's response to images of limousines (and city cars) to sports cars. Indeed, images of sports cars more readily activated brain regions receiving dopamine input or those sending information to dopamine neurons—the ventral striatum, the nucleus accumbens, and the medial orbitofrontal cortex. The results were plain. In young male brains, sports cars clearly engage reward prediction circuitry more than the other tested car types, suggesting a neural correlate to what we already know from behavior—young men like sports cars. They will earn money, rearrange their lives, and altogether pine over sports cars. This experiment was a first step, but an important one; our culture has assigned social status to sports cars, and this status generates measurable responses in brains. These kinds of experimental probes open up many new types of questions relating brain responses to cultural messages. This is where we turn to the cultural street fight, the Pepsi Challenge. Perhaps a similar experiment using simply the "idea of Coke" and the "idea of Pepsi" might also reveal comprehensible brain responses?

Sam McClure, working with me as a graduate student, addressed just these issues. Together, we carried out the Pepsi Challenge but with a twist: We scanned the brains of volunteers to uncover the neural responses correlated with "brand image," not just the brain response to the contents—the caffeinated sodas. The question asked was simple: Did the mere idea of a brand engage reward-processing and -harvesting mechanisms housed in our midbrain, striatum, and orbitofrontal cortex? Studying the brain's response to reward using functional MRI has now grown into a vast scientific industry; however, to tease out brain response to "brand image" required a slightly different style of experiment. A direct reward experiment can examine brain responses to the delivery of substances like sugar water, caffeinated drink, ice cream, and so forth. The challenge in these approaches is to make sure the measured responses are due to the rewarding aspect of each stimulus and not something less interesting like brain responses correlating with swallowing (e.g., sugar water) or merely reflecting the temperature (e.g., the ice cream).

McClure sidestepped these issues in two ways. First, he didn't just measure brain responses to Coke and Pepsi delivery; in fact, he removed this effect entirely. Instead, he administered the Pepsi Challenge outside the scanner to reveal the drink preferences of the subjects—that is, what they chose. Behavioral economists call this revealed preference; subjects' behavior reveals their otherwise private valuation of the sodas. He then delivered Coke and Pepsi to subjects while they lay inside the scanner, but passively—he just asked them to look at a screen, a light would come on and go off, and six seconds later, a cola would be delivered. This simple conditioning experiment was invented by Ivan Pavlov, the Russian physiologist and behaviorist. He was the first to characterize clearly what in previous chapters I have called value learning. Pavlov is famous for his landmark learning studies, now called Pavlovian or classical-conditioning experiments. In Pavlov's experiments, hungry dogs would be exposed to a neutral cue, like a bell. The experimenter would ring the bell, wait a bit, and then feed the dogs. At first, the dogs would salivate at the time they saw or smelled the food. But by the association of the bell with food, the salivation response transferred to the bell that predicted food. The Coke-Pepsi experiment was just like Pavlov's: Ring the bell, wait a bit, feed the dog. Repeat. McClure's arrangement was just the same: Turn light on/off, wait a bit, squirt, and subject swallows. Repeat. It was like some kind of dental office nightmare, but effective. But there was a twist—two twists, in fact.

McClure was not interested in a subject's brain response to each drink; he already knew that water, sugar water, and sugared sodas all cause almost indistinguishable brain responses if no other information is given to subjects. Instead, he wanted to identify brain responses that *predicted* each subject's choice made earlier outside the scanner. There is a big difference between these questions. Looking for brain responses to a cola squirt is at the front end of a series of events that lead to a reward-guided choice. Looking for brain responses that correlate with choice is at the back end:

cola squirt → evaluate the cola → decision to select one cola

Many experiments have evaluated direct brain responses to rewarding stimuli like a cola squirt, a beautiful face, art, drugs, and so on. These experiments probe the left end of this chain of events—uncovering brain responses to each rewarding stimulus. The results have identified common brain responses to a vast array of rewarding stimuli—images of faces, drinks, food, money won in gambling games. Consequently, we now know that the "rewarding" aspects of stimuli and behavioral acts are likely to be computed by common brain regions. But McClure's experiment probed the right-hand side of this chain of events. Instead of measuring the response to squirts of each cola type, he measured the brain response correlating with their choice, but in a situation (passively drinking cola squirts) where no choice needed to be made. Since no choice was made inside the scanner, his experiment was probing each brain's *valuation* of each cola type—the "evaluate the cola" step.

It's important that drink delivery inside the scanner was completely passive, that subjects only be told to swallow the drink squirts. For example, if subjects had been told that "later we will ask you to evaluate the liquids and tell us how much you like them," then all sorts of other brain activity might have occurred related to this "prospective evaluation," that is, evaluation with the intent to use them in future decisions. The same reasoning applies to making choices inside the scanner—McClure would never have been able to separate activity related to valuation and activity related to movements or movement plans. His trick revealed strong activation in a region called the ventromedial prefrontal cortex, a region just above and between your eyes (but of course inside your skull).

As I mentioned in an earlier chapter, in Antonio Damasio's *Descartes' Error,* the ventromedial prefrontal cortex was exactly the region lost by the nineteenth-century railroad foreman Phineas Gage when, in a dreadful accident, a bolt of iron shot through his head. The changes in Gage's personality have been described many times—but Damasio and his colleagues have shown that this region is essential for normal decision-making. This is exactly the region identified by McClure as being essential in the valuation step for choosing (not simply responding to) either Coke or Pepsi.

Now for the second twist. When the Pepsi Challenge was performed outside the scanner, there was absolutely no relationship found between the colas picked and the colas that subjects bought in stores! The finding was so remarkable that the experiment was repeated over and over again with a large number of subjects. Same answer. If subjects did not have any knowledge of the brand, then they apparently chose at random relative to what they normally traded their money for in stores. Another researcher helping with the experiment wondered whether the "brands might taste like something." I've argued earlier that we already should suspect this is true based on what we know about the dopamine system. To test this possibility a twist was inserted into the experiment. All pairs of cups were filled with Coke, but only half were labeled "Coke" (by hand). Subjects were told that the labels were accurate, but that the unlabeled cup could contain either Coke or Pepsi, which was true. In this case, the *contents* of both cups were exactly equivalent and the only differentiating feature was the word *Coke* on one cup in each pair. Subjects overwhelmingly chose the "Coke"-labeled cup. Initially, this could have been simply due to uncertainty—like our example of the Coke machine. Subjects were merely avoiding the uncertainties of the unknown cup. But not so—an exact duplicate experiment using Pepsi and labeling half the cups "Pepsi" yielded no effect of the brand name. Without going through details, the brain imaging results confirmed the behavior. Coke was delivered on all trials in the scanner and the only difference was subjects' knowledge of whether it was Coke being delivered. No effect was seen for Pepsi. Several more manipulations reconfirmed the results—the Coke brand had a flavor, or at least was a major contributor to the experienced flavor. Not so for the Pepsi brand.

These experiments show clearly that a cultural message, the brand image of Pepsi or Coke, has differential representation in people's nervous systems in such a way that this brand knowledge can be visualized in fMRI experiments, where its influence on choice can also be measured. But why Coke and not Pepsi? The results of the first Pepsi Challenge imaging experiment do not tell us why one message burrowed into subjects' nervous systems and the other did not, but it does open up the possibility of understanding conditions under which messages control decision or valuation machinery in the brain. It provides an excellent adjunct to behavioral assessments of the influence of messages. More importantly, it gives us a tool to probe how a malevolent message like "go to the next level" became so potent in the nervous systems of the Heaven's Gate members that it vetoed their every instinct for survival.

Perhaps this sanguine assessment is premature. Maybe the responses in the ventromedial prefrontal cortex and other reward-processing regions were somehow tied specifically to drinks, or perhaps just drinks with caffeine and sugar—or some other scientific "gotcha." Caffeine acts directly on the dopamine system. It prolongs dopamine signals in the brain and it increases general arousal. Although the "brand image" experiment controlled for this direct effect of caffeine, it was still possible that caffeine was needed as some kind of "trigger" to make the experiment work. However, several groups have now moved from primary rewards like sugar and caffeine to a stimulus that is rewarding, sometimes deeply rewarding, but several levels above the cultural skirmish taking place in the soda machines: art.

Hideaki Kawabata and the neurobiologist Semir Zeki have addressed directly the human brain's response to art by asking subjects placed in functional MRI machines to rank art as "ugly," "neutral," or "beautiful." They saw a variety of responses in visual areas, just as expected. They also observed activity in the motor cortex and another region, called the anterior cingulate cortex, known to monitor "output conflicts," which are just what they sound like: opposing outputs competing to control our singular choices. All these findings were consistent with the fact that subjects had to evaluate a painting, judge its beauty, and make an action to report their decision. The most pronounced activation that scaled directly with each subject's beauty rating was in the same region of the orbitofrontal cortex that McClure and colleagues identified as reflecting a valuation response that predicted subjects' choice of cola. The overlap between the Coke-Pepsi valuation responses and the art-beauty rating responses was remarkable; however, the experiments were different in an important way.

Kawabata and Zeki had subjects making choices in the scanner, and each choice was associated with movement. Their experiment was designed to answer questions about aesthetic judgments, not just the valuation part. In the Kawabata and Zeki experiments, aesthetic judgment is a complex construct since it includes valuation, planning to act, and acting (moving to indicate rating). But the overlap was provocative: valuation responses in common brain regions for both art preference and cola preference. Perhaps the ventromedial prefrontal cortex was a central player in computing the relative value of all stimuli ranging from the value of soda to the value of experiencing the *Mona Lisa*? Paul Glimcher, in his book on neuroeconomics, would strongly support the idea that the brain must place diverse stimuli on common valuation scales. If we also include twenty years of work in this brain region by the Damasio group, a consistent picture emerges—the medial orbitofrontal cortex appears to integrate an extremely diverse range of information sources to compute common internal currencies—common valuation scales. Phineas Gage lost an extremely important region of his brain.

Inspired by this collection of findings, Ron Fisher and Ann Harvey repeated the McClure paradigm for probing brain responses that relate directly to a subject's preferred choice. But instead of using colas, they used masterpieces of Western art. The idea was simple: Put people in the scanner and ask them to look at paintings that are displayed at unexpected times. That's it—just ask people to look at the paintings, no movement, and no preparation for future evaluations of the paintings. Just look. Once this was complete and subjects exited the scanner, they were asked to evaluate the paintings for "liking" and "familiarity." The second step was voluntary, but most subjects agreed. The outside-the-scanner assessments acted just like the Pepsi Challenge tests for the cola task by producing numerical rankings of each painting for each subject. Humans respond quite idiosyncratically to art; one person's beloved painting can cause revulsion in someone else. The same holds for soft drinks, foods, and just about anything else that requires aesthetic judgment. So Fisher and Harvey extracted personal ratings from each subject and then used these to query the brain responses measured during the passive viewing of the paintings. This move identifies brain responses correlating with their idiosyncratic valuation of the collection of paintings. They immediately hit the jackpot—a strong response emerged right in the same place as the Coke-Pepsi preference response. Also, there was no motor cortex involvement, since no movements or even planned movements were required in this task—only passive viewing of paintings. Fisher and Harvey found something else as well—in this valuation region of the brain, familiarity and liking are not represented as separate variables. They found the same correlation behaviorally—subjects liked paintings with which they were familiar.

EVERYTHING WE EXPERIENCE is a composite, made of lots of different mental parts; even what we like to eat and drink. When you eat carrots or chocolate or even drink soda, your experience, what you taste and like, is not just a response to "content." And we've just heard how even the idea of a brand has an impact on flavor preference—that is, on valuation. But why should the brain invest itself in valuing proxies this way? Again, it's about efficient learning; these odd examples that we're examining show symptoms of our valuation mechanisms operating in modern culture. The problem is that modern culture presents all sorts of conflicting messages that our nervous systems absorb, value, and then use to make choices. And although we think some choice will make us happier or more satisfied, we often make choices that degrade our lives instead of upgrading them. So why do some messages insinuate themselves into our nervous system to commandeer behavior while others do not?

This important question raises social, scientific, political, and medical issues. In earlier chapters, we made some headway at understanding broadly the underpinnings of why certain messages might gain behavioral power. They come to act like rewards, and the rest of the brain adapts itself to predict and acquire them. Events that foreshadow these potent messages also accrue value because our brains are designed to transfer value to events that predict reward. Just like "A friend of my friend is also my friend," the brain has its own version, "A predictor of a predictor is also a predictor," where the predictors predict future reward. This is exactly why even complex verbal descriptions like the ". . . salad of perfectly grilled woodsy-flavored calamari . . ." can set off our reward-seeking circuits. It's a proxy for the reward to come. Yet none of these observations tells us what makes one message more efficient at garnering "reward status" than another. This is a question about the assignment of relative value.

Let's consider again what all this proxy machinery is doing in our brains—it's all about advertising and the fact that nothing we experience is really what it seems to be on the surface. That's right; the proxy machinery is the ultimate marketer. In the real world, marketing is that business of bringing buyers and sellers together. In principle, it's a lubricant to making a market work. But where did marketing begin?

Marketing and advertising are not innovations of modern civilization; they are deeply biological functions. For example, flowering species "understand" this problem well. Flowers market nectar (sugar water) to bees; the bees consume the nectar and thereby help the flowers to pollinate and reproduce. Although many plants self-pollinate, there are plants where bees represent the major agent of pollination and so are a natural extension of the plant's sexual cycle. The sad fact (for the competing plants) is that most flowers market the same basic products, sugar water and pollen, to bees and other insects. They are all shackled to hawking the same elixir—don't look for fairness in your garden. Although this sounds like an impossibly tough market, the flowers and their attendant bees have struck an alliance. This alliance is an old one, and flowers have evolved complex advertising campaigns all directed at tickling the nervous system of the bee and compelling it to land, brush around on the plant's sexual parts, and fly away to do the same to another plant of the species— presto, pollination. This is a symbiotic scheme—the bee needs energy and amino acids, and the plant offers them in the form of sugar water (energy source) and pollen (protein source). Consequently, the bee's nervous system comes preequipped to "like" pollen and nectar, and the learning machinery in its brain is equipped to assign value to other stimuli that predict either of the prewired rewards. Cross-pollinating plants need an agent to help them reproduce—literally carry their male gametes some distance—and the bee delivers in the form of an efficient gamete shuttle, carrying pollen grains around from plant to plant.

Plants and animals have been advertising for as long as there have been plants and animals. And their survival depends on effectively selling an array of messages. "Don't eat me, I'm poisonous," "I'm likely to produce strong offspring," "I'm very healthy and this is why I'm able to produce this lush, richly colored tail," "That scent means that I'm mad." Since an organism cannot verbally announce its fecundity or ferocity, it must employ lower-cost external proxies for its internal propensities. Such advertisements are meant to be understood as representations of important, more complex internal qualities.

Berries are another example of multimedia advertising and distribution. Plants that make berries wrap their seed in carbohydrate; this starchy jacket is further wrapped in a package that is bright and easy to see when ripe. Berries are often packed with vitamins. So a purchasing bird is better off for eating the berry and the plant gets its seed spread. An efficient market when it works. Of course, the plant's "choice" of berry properties is tuned to capabilities of the nervous systems of their clientele. The "hidden" seed of the plant is typically tough enough to make it through the digestive tract of the bird, and in some cases, the seeds taste bad, causing the birds to spit them out. How dare these plants use such tricks! They are deceiving the birds just to get help copying themselves into the next generation.

This effective messaging problem lies at the heart of an important domain—the use of terrorism as a political tool. You must wonder—surely terrorism as a strategy depends on all sorts of cultural factors and historical trajectories of different groups and so on. Sure, of course it does. But to my mind, the deep problem with terrorism is encoded in a shallow observation: It works. In a certain way, it works wonderfully, and it is much more effective now than it would have been a hundred years ago because its impact depends on the fluid, fast communication channels now provided to the entire world at very little cost—broadcast television, cable television, Web sites, instant messaging, chat rooms, cell phones, and so forth. So depictions of acts of violence—exploded buses, demolished restaurants, burning pipelines—although many thousands of miles away, spread out all over the globe almost instantaneously to reach our brains, where they propagate through our neural networks, which assign value to the "idea of terrorism" or the "threat of terrorism." Of course, our brains will also pick out proxies for the terrible outcomes these "acts of terrorism" embody, and this value may propagate to other proxies. It doesn't matter whether someone's personal likelihood of being harmed has measurably increased; the value passing goes on and the brain damage is done—our brains pick out symbols that represent these terrible outcomes and overvalue them, and this confers inordinate power on the mere idea of terrorism. So just as the idea of Coke or Pepsi has representation in your nervous system, so, too, does the idea of a terrorist attack. We have a lot of behavioral evidence that these are very influential proxies. For example, no one is sure how to keep from overreacting to these "possible threats." In fact, it's extremely difficult to decide what an overreaction might look like—exactly who is willing to risk the life of friend, family, neighbor, countryman, etc.

by not overreacting? Not me. And within our nervous systems, the proxies for these terrible acts multiply. The value (here negative value) of terrorist acts has passed from one symbol to the next, creating a literal cornucopia of terrorism predictors in our brains—a real impediment to clearheaded decision-making about these issues. How many different cues make you think of terrorism, if only briefly? A grenade? Broken tempered glass? A man with black hair and a mustache? An airline ticket? A folder of matches? How many of them seem like an overreaction?

Can our new knowledge about valuation in the brain and its connection to real neural systems shed light on how the messages of terrorism, the cultural messages, are processed and prioritized by our minds? I'm not sure, but nobody has really tried yet. The fact that reinforcement learning models can give insights into dopamine-guided choice in humans has already inspired funding agencies to wonder about using the models to understand drug and gambling addictions.

Bill Casebeer, the air force officer I mentioned in the last chapter, believes we can use brain imaging paired with behavioral experiments to understand the features that make one cultural message more potent than another. He wants to understand two things at once: (1) How do the cultural factors that encourage someone's reception of an inciting message—a message that could incite the receiver to violence—influence brain responses? And (2) what brain systems are involved in each step of this process? Casebeer's guess is that the reinforcement learning systems in the brain must play some role, particularly since they are powerful enough to direct drug addictions. They can make someone trade her family, possessions, and even her life in order to put white powder up her nose or inject a fluid into her veins. These same systems can also be hijacked by ideology, just like the Heaven's Gate cult. It seems reasonable to begin to apply our current models of value learning to these important problems.

8

OUR CHOICE

It's Not Your Mother's Soul, but It's Still Alive

WITH THE PUBLICATION OF *The Origin of Species* in 1859, Charles Darwin demystified diversity in the biological world by exposing the most powerful learning algorithm on this planet—natural selection. Variation lets biological systems explore alternative solutions, selection provides the feedback, and storage allows the system (the natural world) to retain the solutions that worked. The backdrop for this algorithm is the real world, where the harsh demands of survival set standards for what constitutes successful solutions. Through repeated application of these steps, grand complexity accrues to produce adaptations of spectacular beauty. Every biologist uses these ideas to think about the mysteries of the living world. It was a stunning insight. People had looked at the same world, the same data, for millennia, and only one person saw a nonmysterious explanation for its awesome diversity. Darwin spent almost twenty years pondering the ramifications of his theory before publicizing his ideas, but that first step, the notion of natural selection, was clear and singular.

Almost a hundred years later, Alan Turing recognized that any process that could be described was itself a computation, a processing of information. His ideas went two big steps beyond Darwin's. First, Turing's ideas transform Darwin's discovery into a computation—they portray evolution itself as an algorithm. Vary—select—retain or discard. Turing's ideas show that evolution is a computational process and it is computing organisms. The computations are running on dirty, goopy, slimy components, but evolution is an algorithm. Perhaps the ultimate algorithm, whose output includes building conscious humans, the kind that can veto their instincts in pursuit of an idea. But Turing's ideas take another big step. They unite life, mind, and machine by grounding the idea of vitalism in the same way that they took care of the mind. The idea that life's vital essence and mind's vital essence is indescribable and comes to us from some unphysical place was fashionable in the late nineteenth and early twentieth centuries—so fashionable that few contenders existed to counterbalance prevailing views. But Turing's key idea about computation grounds both using a single idea.

The "living" that a cell does is the information processing supported by its parts and their dynamic interactions, and the "thinking and feeling" that a mind does is the information processing supported by the brain's parts and their interactions. Turing stepped beyond Darwin in revealing life and mind to be computations. Life and mind present themselves along an axis of computation that stretches from simple forms of life to mental operations of sophisticated mammals. But this poses an interesting challenge yet to be answered. Plants are living but aren't considered to be thinking. People are

living and thinking, equipped with a behavioral power to veto their
instincts. Here's the rub. It's a crime to harm a human, but trimming
a tomato plant is not. This intuition seems obvious to those of us who
are not Buddhists, but we need a means to distinguish the patterns
of computations that constitute a tomato plant from the patterns of
computations that represent us. And then there are the incredible
number of "in-betweens," the vast spectrum of possibilities ranging
from the tomato to a person. If we take Turing's ideas seriously, then
we will be driven to develop methods to characterize computational
complexity with more much finesse in order to distinguish the person
from the tomato—current methods are far too blunt to deal with this
problem. And even when such quantitative assessments become available, the use of such assessments will still reside in our social institutions. Science will not remove the need for us to think responsibly
through what we want; it's still our choice.

THE IDEA OF ASSIGNING value to computational characterizations
of life and mind raises a specter. I dedicated this book to explaining
how a computational system can care, that is, how it can assign meaning to apparently meaningless streams of symbols. I also argued that
this way of posing the question was inappropriate. As the evolution
algorithm built ever more complex computations for generating behaviors and thoughts (ever more complex minds), these patterns of
computation already carried with them measures of their value back
to the organism—their meaning. Iterated over eons, this process built
us, but built us out of parts already endowed with a measure of their
meaning. This is how evolved value can be embedded throughout our
bodies and why it confers meaning to the information-processing

running on our brains. Only one area of valuation received any detailed discussion—reward-dependent guidance signals. There are many other behavioral and mental domains where valuation mechanisms are not understood, and I fully expect this to be a growth area in the near future. But the unsettling point, and perhaps the one not to write home about, derives from the unnerving message about meaning at the heart of my book: All meaning is physical.

The computational theory of mind began with Turing, and a computer revolution emerged from his efforts. The idea accounted for "mindlike stuff" in terms of interactions of "stufflike stuff." Transcendent step, but meaning was missing—there was no provision for one computation to be assigned more value than any other. Turing's idea left something out. One stream of symbols was as good as another, so a computational device in this view could only transform symbols into symbols. The "bombes" built by Turing to break the Enigma codes used this model of computation because they were oversupplied with energy—they never had to assign deep meaning to one computation over another; they weren't forced to. Energy was like air, and it was oversupplied because winning the war was so urgent. This was a misleading microcosm—a kind of historical hiccup. For only in a world with unlimited energy could such a model of computation evolve—this is not our world, and so this is not the solution selected by the evolution algorithm.

So, what gives meaning to biological computations? Valuation is meaning, and valuation arose because of costs. Costs forced on all biological computations the need to be valued. Living systems all run on batteries, energy is limited, and life is desperately hard. This is why choice and the ability to choose evolved. Evolution selected for organisms that could value their computations wisely; not just momentary valuations, but deep valuations. Deep valuations are those that compile the long wisdom of past experience with the best guesses at the likely future. This is what evolution should have selected. I'm guessing that it did. However, I'm going to take another step. If the valuations build in guesses about the likely future—what might happen tomorrow—then why restrict the guesses to tomorrow? Why not have valuation functions that include information from the experienced past with all the other possibilities: "what might happen" in the future and "what might have happened" in the past? Why not include all possible pasts and their evaluations and all possible futures and their evaluations? No reason not to—so let's do it. Good valuations of the past should include guesses at the outcomes from past choices that were not made as well as those that were made. Evolution "noticed" this as well. The claim finds support in the facts discussed in Chapter 6: We have the word *regret* in our lexicon, in our observed behavior, and in our measurable brain responses. Natural selection chose individuals with decision-making machinery that took account of what "could have been" in the past, present, and future.

This is why biological computations are so crafty—they can profit from the entire history of life and compile in life's best guesses at the likely future and the pasts that "could have been." Take the example of Deep Blue (the chess-playing program)—it used a stored history of past chess games to form valuations of contemplated moves that included the future value of making the move. So its evaluation function possessed wisdom of the likely chess moves to follow and whether they would give or yield advantage in the future. I'm saying that valuations in organisms have no principled reason to be tied only to the experienced past and possible future. Instead, they should include the experienced past, the experienced present, the likely future, and even the likely past. They should include broadly "what might happen." I'm not surprised that languages include the subjunctive mood; our bodies do. I'm also not surprised that our physical theories also include it.

Deep valuation solves immediately another problem: It can guide a moment-to-moment resource distribution *strategy*. If the valuations are correct, then investing energy in proportion to value will yield better reproductive success overall for the organism. Two problems get solved at once: Meaning is attached to computations in proportion to their value, and the valuations direct how energy should be efficiently distributed. It's a remarkable trick—at any moment the valuations are the strategy for energy distribution. It's like an optimal battery-draining algorithm. Think about it—it's really a principle for coupling computations (with their valuations) to energy use. It's easier to think about for single-celled organisms. They should distribute energy to their internal processes according to the relative value of those processes. It's not enough to know how to distribute resources correctly. An organism must actually do it. I would expect, therefore, an intimate link between valuation and the capacity to act on it as a strategy for distributing resources.

There is a fundamental flaw in the old story of the computational theory of mind, the one that assumed that mental operations could ever have evolved without any meaning attached to them. It's almost impossible to imagine meaningless mental operations evolving in a world with limited energy and equipped with the first law of efficient computation: Recharge or die. The first step in the old story left something out—batteries. Computations cost, and so the information-processing solutions that evolved were built out of parts and dynamic interactions already equipped with a way to measure their meaning to the organism. The meaning was always there; it's one of the features on which evolution acted.

So what are we humans? We are meaningful patterns of information processing composed of two kinds of computation: those computations on which all perception and movement depend plus their valuations. In biology, these two computations are bound together in an intimate embrace because it's senseless to have one without the other. It's worthless for the system to possess a computation whose long-term value is not known; how would the system ever choose to use it? It wouldn't because it could not afford to risk energy on it. That would be like gambling with absolutely no chance of winning—none, not even a really small promise. Like the unknown brushed-aluminum soda can without a label, we don't trade our money for it because it makes no promises at all about the payoff of its contents. In fact, we're not sure if it has contents.

If life were plugged into some magical electrical socket delivering unlimited energy, our brain's mechanisms for choosing would be different. Choice might not even exist. But such magic plugs do not exist and we did not arise in that world. So choice is about relative value, and relative valuation arose because life runs on batteries, energy is limited, and there's no free lunch. Energy must be distributed to computations according to their expected payoff. And this is how biological systems choose, this is how we choose, and this is why biological computations are freakishly efficient.

Energy distribution systems in the brain know exactly how much energy to commit to a computation: an amount proportional to its current value. This is why my forehead, your forehead, and even Garry Kasparov's forehead are merely warm to the touch. All biological computations are paired with value computations, and this is why biological computations can care. But there is more. Patterns of biological computation come fully packaged as whole creatures—moving this way or that—chasing some expected returns and forgoing others. In such systems, control over the whole package is required. The whole package must have a way to set and pursue goals, and possibly innovative ones. This is why our own guidance signals, broadcast from the deepest recesses of our midbrain, can bestow to even our most abstract thoughts the power to guide behavior the way food and sex can. This is why human ideas can veto survival instincts. Sharks don't go on hunger strikes because they can't form and maintain ideas as we can. If sharks had prefrontal cortices like ours, they, too, might go on hunger strikes to protest overfishing of their kind or even try to hook up with spaceships on the other side of comet tails. I saw the movie *Jaws* as a fifteen-year-old, and so I must admit that I'm glad sharks can't form cults; they are scary enough already.

WE BEGAN WITH TWO simple questions that task every creature on this planet: Which choices are worth making and how much does each cost? These questions identify two partners that together shaped the character of every psychological mechanism that inhabits our head: value and cost. Psychological mechanisms are as diverse as the problems they solve, but value and cost have been the guiding hands operating in the background since life evolved. Given the significance of value and cost, the key principle of efficient computation becomes obvious and has acted as our guide throughout these chapters. The principle of efficient computation plotted a rising course from neural properties through goal-directed learning and upward to trust and regret. And from there to where? Are we to feel good or bad about gaining scientific insight into trust, regret, and choice? Author Tom Wolfe portrayed his reaction to the sister sciences of mind and brain:

> Brain imaging was invented for medical diagnosis. But its far greater importance is that it may very well confirm, in ways too precise to be disputed, certain theories about "the mind," "the self," "the soul," and "free will" that are already devoutly believed in by scholars in what is now the hottest field in the academic world, neuroscience. Granted, all those skeptical quotation marks are enough to put anybody on the qui vive right away, but Ultimate Skepticism is part of the brilliance of the dawn I have promised. . . .
>
> Neuroscience, the science of the brain and the central nervous system, is on the threshold of a unified theory that will have an impact as powerful as that of Darwinism a hundred years ago.

Powerful . . . Wolfe is concerned about whether we will lose ourselves when these conceptions of self are closely scrutinized with imaging devices, genetic probes, and computational models. Will these ideas about what guides our behavior from moment to moment change us as they fall prey to the "unified theory of neuroscience"? Wolfe has been misled a bit by rumors of our soul's death, rumors of the impending publication of unified theories of neuroscience, and a misleading metaphor about what we are.

Physics, already 150 years ahead of neuroscience and fully outfitted with powerful formal models, should be so lucky as to have a unified, agreed-upon theory of how the universe works, but it doesn't yet. Theoretical and computational neuroscience is embryonic by

physics' standards. Theoretical neuroscience is still a loose confedera-
tion of computational models, some quantitative yet restricted to specific domains, but most semiquantitative, making predictions that
agree quantitatively with some data and qualitatively with the rest.
However, don't be fooled by this circumspection. It's just my report
of the current state of affairs, the current position of the field; the
speedometer readout is missing from this report. The march of these
models is forward not backward, real progress is being made on
many fronts, and things are indeed accelerating. Also, we need not
cower at the notion that neuroscience is about to be "explained" nor
should we fear its products in the coming years. And the bit about the
soul—while we intuitively know what Wolfe is suggesting, it casts the
problem in terms that our mothers would understand, but our mothers, for all their innate wisdom, did not know enough about the brain.

The soul is not dead, but it's also not your mother's soul—it never
was. Let us for a moment address that older notion of soul, the one
whose death Wolfe has proclaimed, the one given to many of us in
our youth by others who cared about us or simply told us what they
knew. As we have discussed before, there may exist features of reality,
unknown to us now, that exert influences on our bodies, brains, and
minds. There is no real downside to admitting the possibility that science is incomplete and just moving on, but it's often the moving-on
part that is difficult for some. However, science, as a method of knowing the world, is about empirical inquiry and explanation using observable or describable entities. If you rule out physical explanations,
then useful discussion ends. In my description here, my mother's soul
is a meaningful pattern of computation, built out of parts that value
("care about") the fate of the whole organism—not an indescribable
essence injected into our bodies at some point during our personal

character arcs from fertilization to grave. So, it's not my mother's soul. But even the older ideas about the soul share one important quality with these newer ideas: They are ideas. Therefore, they must run on our brains.

Ideas about the "soul" or "me" or "self" or "immaterial force" accrue value in the way that I have described, and they can and do motivate behavior. "Ideas about the soul" are cherished by many of us, and we modify our behavior based on them. Ignoring this pattern of behavior in humans or simply labeling it as illusion would be unreasonable and certainly unproductive. If it's an illusion—an effect of running up against the limit of some operating range— then we must ask what's the quantity whose meter is constantly pegged at some end of its design specs? If it's an illusion, then I propose here to call this "bug" a "feature" of mental lives. I'm giving it airtime in this chapter, and simply by virtue of that it gains a presence in your cortex. Your brain must now decide how to react to the discussion, how to categorize my arguments about the soul, and maybe even to revise your ideas a bit about how brains could have souls—at least the new kind. "Ideas about the soul" also generate reproducible patterns of behavior chronicled in every culture, they must be represented in our brains, and they can gain value. Our mother's soul, the one without a description and coming from somewhere else that can't be identified, can't be the one running our brains. But the "idea of the soul" is certainly there, humming along and influencing behavior.

THESE POINTS MAY SOUND ACADEMIC (I'm officially considered an academic), but the discussion of who we are and what we think we are has resounding practical implications. What if we trim down our modern notion of the soul to merely our capacity for agency, our ability to choose? With this trimming, can we make some progress in understanding the nature of the agent that makes choices? Daniel Dennett addresses these questions in *Freedom Evolves.* He makes a cogent case for the ways in which we are free to choose and how these capacities evolved. But between evolutionary arguments and behavior lie specific, proximate brain mechanisms. How do we sustain a capacity for choice in the presence of brain mechanisms that constantly seek to make choices, thoughts, etc. efficient and automatic? If ideas can veto instincts, what happens when the ideas become automated and habitual, and pretty much insensitive to new data? How does our capacity to choose deliberatively remain sensitive to new information? You see the trade-off. Our brains seek to make representations of ideas and actions, and sequences of both, efficient. There is a pressure for the deliberative to transition to the automatic, transforming the thoughtful into the thoughtless.

Brains that followed the efficiency prescription were like the Energizer Bunny; they propagated better than the rest. In earlier chapters, I focused on reward-prediction systems and showed how value propagates from one proxy to the next, from predictor of reward to a predictor of that predictor of reward and so on. But the general idea hiding behind these details is that the brain is seeking to convert the deliberative into the automatic or habitual—converting effortful thought to a habit of thought because that's efficient. Novel ideas (innovations) start out in our prefrontal cortex, require attention and effort to sustain and follow, and can be updated by new information flowing into our brains. However, as soon as an idea is plugged into the "reward slot," the rest of the brain tries desperately to find efficient representations that guide acquisition of this "idea reward." If this "idea reward" helps chain several actions (real or imagined) together, then eventually this entire chain of internal states will be represented by a single proxy, a single symbol that stands for the entire sequence. Imagine, or remember, learning to serve a tennis ball.

At first, you learn the component parts of the move. How to hold the racket, where to look, how to hold and toss the ball, and how to stand, follow through, and end up in a stance that's ready for your opponent's return. Early on these components are learned separately. Later, the separate components are linked together into groups, and finally the groups are linked into what one hopes will be the ballet of a perfect serve. One action chained to another chained to another— then, as learning occurs, these chains of actions become more and more automatic until finally one becomes quite adept at treating the entire chain as a single entity, a whole. At that point, thoughts actually intervene to mess up the entire process. Focusing attention on one part of the serve now becomes an impediment to serving correctly. In the beginning, this focus was required to learn the parts and to link the parts together, but once this is done, the entire act becomes like a single object, like a habit. Any athletic coach will attest to some similar story. Now suppose this was a problem in arithmetic to add fractions together—you are taught one part of the process, then another. Eventually the entire act will become automatic. Later on, you may even hesitate when someone asks you the rule, but still be able to add fractions quickly. The same features attach to legal arguments, running cash registers, reciting poetry, or just thinking through some class of problem that faces you frequently. We can observe this automatizing of actions in our own behavior, and such chains of behaviors become automatic, aided by numerous mechanisms including the value-passing scheme discussed in Chapter 4. If the chain of behavior has acquired a "reward" (like a compliment or a juice squirt) over and over again, then removing the need for component-by-component focus saves time and energy. But what's happening in the brain?

The learning begins with a dialogue mainly between your prefrontal cortex, striatum, and midbrain using the reward-harvesting mechanisms there. The learning also requires short- and long-term memory, and so those structures are also involved. As the chain of behavioral components (or ideas) becomes more automatic, it changes status; it becomes a multipart script all summarized under a single symbol—it's like a button. Now the entire serve is a "serve button" in our brains. And unless our serve is going very badly, the serve button shouldn't be cracked open and its internal parts disturbed. That would be inefficient—a complex device (and a chain of behavior is a device) should not be opened up to fix it unless there is no other solution. The reason is that there are innumerable ways to mess things up. We all know what happens to any complicated motor act when we worry about one part of it or become distracted—something goes wrong; the system makes mistakes. The brain seeks to change the status of a guiding idea ("idea reward") from a data-sensitive, deliberative guide to a sleek, automated guide, one that doesn't have to check with reality nearly as often. In some contexts, this kind of distinction is called implicit versus explicit knowledge, but here we're mainly talking about behavioral control—so we're talking about the transition from deliberative (explicit) to automatic (implicit) behavioral control. It's reasonable to suspect that this transition correlates with a transition of neural activation from the prefrontal cortex to the striatum, and experiments have already clearly delineated goal-directed learning systems from habit-learning systems. One innovates and the other helps transition the innovations from their early clunky stage to efficient schemes for thought and action. But there is a gap in this sequence. If the brain seeks this transformation to thoughtless efficiency, why don't all ideas eventually end up in that state? And if they are all driven to the state of "thought habit," how can "we" remain intact? How can we ever break the "rules" that our habits of thought have laid down?

THERE IS AN EPILOGUE to the story of my uncle, the one who valued so highly getting to the Georgia–Georgia Tech football game. And it makes a point about breaking rules.

After driving rather carefully down the sidewalk to our cross street, we then zoomed on to a building near the stadium, but possessing a guarded parking lot just behind it, fully outfitted with a little guardhouse and a crossbar to stop cars. As we drove up to the crossbar a guard emerged with half a sandwich in his hand and the rest in his bulging cheek.

"I'm sorry, sir, the campus is closed today and no one can use this lot." I could barely make out the consonants, but was thoroughly captivated by the movement of his swollen cheek.

The guard leaned against the car with his free arm and maintained a friendly face.

My uncle grinned at him (and I think at me first) and said, "That's OK, I work here. I have a faculty parking sticker."

A fluttering half-blink danced across the guard's eyelids—one of those "here we go again" gestures.

"Sorry, sir, but this lot is closed as tight as ———— to everybody but God and the governor of Georgia."

Where do they get these guys?

My uncle's head dropped slightly. He inhaled again. That was my "oh, no, something bad is coming" cue. In response to my heightened state during the sidewalk-driving bit, my brain must have overgeneralized on his first inhalation—the one immediately followed by our sidewalk roadtrip. My prediction mechanisms kicked into high gear—something "not good" was in store. I immediately looked around for anything illegal and within reach of our car. I felt momentary relief—nothing remotely illegal to do.

"Well, OK, son, thanks for warning me. Would you mind lifting the crossbar so I can turn the car around?"

"Sure." The guard walked to the crossbar, lifted it, and just as it crested his shoulders, a screeching sound emerged from our car, reverberating from all directions. It was so loud I thought another car was running into us.

But alas, *we* were the screeching tires. We peeled out and shot under the crossbar. I winced and turned around to see the guard taking the last two or three steps of a guard-jog.

After a couple of minutes I asked, "Why do you do stuff like that?"

My uncle stopped the car and looked me right in the eye.

"Boy, rules aren't always made to be followed. In this case, does it really matter? I mean, that guard will be telling the story of the 'peel-out professor' for years. Now, get over yourself."

When I was a kid, this kind of exchange was considered therapy and therefore soft—almost too soft for real boys.

How did my uncle emerge as an entity capable of choice when this very capacity rests on mechanisms that relentlessly seek efficiency by reducing our deliberative acts to thoughtless symbols? On the surface, my uncle broke rules apparently based on his relative valuation of two events: getting to the game late and the cost of the "fallout" when the dean called him Monday morning to report the guard's complaint. But it is not clear that my uncle retained agency over his behavior and made a willful choice. We don't know about all of the internal automatic directives against which his observable choice was made. How hard was his fight to choose when measured against the covert, automated directives from his midbrain and striatum? If we could observe those internal "demons" and "angels" that his choices fought against, then we might measure how much true "will" he exercised—we might measure his degree of agency.

The idea here is that one measure of his agency is his capacity to deviate from his internal directives—his learned version of thought instincts—those automatic commands like "Don't break rules," "Money is important," "Be just as smart as your sister," etc. If I could see those commands and the value they carry, I would be able to decide on the degree to which my uncle was exercising choice. That comparison would yield an operational measure of agency, a kind of inside-outside comparison, one made possible only recently with brain imaging technologies. It would be a working solution to the mystery of willful choice.

Behavior teaches us that there are humans in whom agency is diminished, absent, or functioning pathologically, and it's often difficult to know how to deal with them. Surely this measures us as a civilization—the degree to which we protect those or speak for those who cannot do so for themselves. For example, children don't make choices the way adults do, and the law recognizes the need to view them as having a diminished capacity to choose. A two-year-old who takes candy from a store is not prosecuted for shoplifting, but I would be (maybe because I would probably rob a Godiva chocolate store). The concept is not limited to children. Persons afflicted with mental disorders are viewed as having diminished capacity. Brain injury and disease cause impaired capacity to choose. There are already many cognitive tests that can help identify capacities missing or blunted in these groups of people, including short-term memory tasks, working memory tasks, and general intelligence measures.

But there is a problem with choice. The diminished capacity to choose is measured using external variables like the time to make a decision, the ability to distinguish items of different value, the ability to link a series of choices together to achieve a goal. There has not been a capacity to know what's happening inside the skull—to view the externally observed choice against the background of internally concealed neural directives. This kind of comparison is needed because assessment of the degree to which a person retains the capacity to choose helps the legal system determine culpability for an act. The judicial system already establishes degrees of agency (ages at which a defendant can be executed or whether one is insane, for example), but "my brain made me do it" defenses will get more sophisticated in the coming years as neuroscience makes real, physical measurements of it.

Antonio Damasio's work in patients with lesions in the orbitofrontal cortex revealed decision-making diseases. Given that there are full-blown pathologies of choice, there must be a spectrum of capacities to choose, and these will almost certainly be influenced by age, context, and other factors. We even saw in Chapter 4 that some of the symptoms of Parkinson's disease were related to a problem dealing with choice, not movement. A recent experiment involving patients in vegetative comas used fMRI observations of the patients to determine whether their brains "recognized" familiar voices "better" than unfamiliar voices. The idea was that this knowledge would help their desperate families make better, but no less heartrending, decisions about their ongoing treatment. Some regions of the patients' brains gave larger responses to familiar sounds than to unfamiliar sounds, but drawing detailed conclusions about their agency from this kind of probe would not be based on a clear understanding of just what the different responses mean for the patients' agency. The result does not tell us whether "they" are "there." This is one reason why measures of agency should be developed and tested in normal subjects under controlled conditions.

Imaging experiments, designed to probe the internal valuation of choices available to a subject, can be used in conjunction with a battery of cognitive tests to develop a measure of the degree to which a vegetative patient retained agency—ultimately whether he or she is still there. Such tests would need to be crafted to specific situations, since it is possible to lose a cognitive capacity in one context and retain it in another. The details of the experiment would depend on the range of choices probed and their character; however, it is now possible to use an imaging probe to eavesdrop on someone's internal value calculations.

The experiments with the passive valuation of art suggest the feasibility of an agency probe—each person had an idiosyncratic valuation of each painting; however, there were brain regions that showed scaled activity according to each individual's rating and not to specific paintings. Their personal subjective valuations (the numerical score they assigned to each painting outside the scanners) was predicted by a measurable brain response in a specific region of the medial prefrontal cortex. Although a simple measure, these experiments and others like them show that neural correlates of abstract, value-laden choice can be measured. The basic idea is to compare the degree to which an externally observed behavior (e.g., which numerical value is assigned to each painting) deviates from the valuation response for each painting. I'm not going to suggest in detail how to do this, but the capacity to choose in a manner not aligned with one's internal valuations (internal directives) is a crude measure of their capacity to willfully choose. The practical uses of such measures include measuring the degree to which any manipulation like a drug or a television commercial or even a behavioral therapy changed one's capacity to choose, to deny one's internal valuations. Although we don't always want to deny our internal urges, you can see how these comparisons yield a measure of the capacity to choose. In the movie *Terminator 3*, the Arnold Schwarzenegger robot was reprogrammed to destroy the person he was originally supposed to protect; his internal directives kept telling him to kill his former ward. But at a crucial junction, some other process kicked in and vetoed his internal directives to let him choose to not kill. This act represents willful choice, and we can now start to measure this capacity. I suspect that many specific kinds of probes might be developed, and so an entire program of research along these lines would be helpful to science and to policy makers. It would certainly help to differentiate the soon-to-proliferate "my brain made me do it" defenses.

The need to understand choice in a more operational way is growing at a rapid rate. Our population is living much longer than ever before due to improved nutrition, control of infectious disease, and a general improvement in how we care for the elderly. But with age comes cognitive change. In addition to the normal organic changes associated with disease processes, age brings another natural change— the accumulation of mental habits. It's simply a natural outcome of mechanisms that seek to automate ideas. This makes for an efficient decision device, but it also has a darker side—it becomes easier and easier to fall prey simply to habits of thought. There is a real pressure to end up in a state where we may act like a literal collection of mental habits—falling back on one mental reflex or another as the context dictates. This pressure grows with age for the same reason "experience" grows with age. Is this destiny or are there ways out of the pressure of our personal histories?

We know why sharks don't go on hunger strikes and why we do, so surely, the instinct-vetoing possibilities can provide at least a perspective on this issue. Our brain, as an efficient learning machine, is driven to seek cognitive innovation. It's restless for new knowledge, new experiences, and new ideas that themselves may act as rewards and guide further, new chains of mental events. Our guidance systems constantly seek efficient proxies for chains of events and ideas. They "want" to turn deliberate control into habitual, automatic control because that's efficient. It's not their fault—we shouldn't think them malevolent—however, this process can feel debilitating. The habitual whispers of past experience attempt to guide us whether by carrot or by stick: "Don't argue with me," "Your mother would be so proud," "That is DEFINITELY GOING on your permanent record." And this list does not include those directives that never reach awareness, the ones we never consciously hear. These whispering correctives never quite go away and they are particular to us, to the lives we have lived and experienced—our "baggage." These "habits of thought," produced by our past but instructing our present, will be with us for good, but it's also fairly clear that the same system that plagues us with these ghosts also provides a way out—just by our understanding of the nature of its operation. We overlay these automated guides with new ideas that we deem important and that can veto the habits, essentially "catching the commands" being uttered from these pesky little oracles. And these overlays of control that I suggested could be used in conjunction with behavioral probes to measure our degree of agency—our capacity to break the rules. Does our new knowledge of choice confer on us increased agency? Does the reading of this book give you increased agency? I think so. Any new idea you form in reading this book could in principle change the way you proctor your internal impulses.

This is an important issue because there is an ever-increasing pressure being exerted by the knowledge spilling out of neuroscience and cognitive science labs around the world. "My brain made me do it" adds to a growing list of biological defenses in courts and in everyday life. Will we excuse our every choice with some reference to genes or the prefrontal cortex or valuation mechanism? No. It is wrong to equate the mechanisms with the lives we choose to build. It's still our choice. And of course your brain made you do it and genes made your brain and all that, but the time has come to move beyond these simple ideas and get more specific. Let's try to understand how and why "we" stay around, despite the news from the laboratories that efficient processes seek to suck us into a mire of mental habits. Let's take on the idea of "free" choice with all that we have discussed so far.

We do have a superpower. The sad story of the Heaven's Gate cult shows that ideas, however strange, can act with the potency of primary reinforcers—actually something even more important than any natural reinforcers. Dopamine systems located in your midbrain, in collaboration with your prefrontal cortex, striatum, and other areas, participate in the computation and storage of values attached to ideas. The central concept was that an idea, formed in your prefrontal cortex, comes to play the role of a special control signal, a reward-dependent guidance signal. We're not sure exactly how this comes about, but experiments (brain and behavioral) show that it does. The concept is that the idea in your prefrontal cortex exploits reward prediction and harvesting machinery in your midbrain and striatum to "trick" the dopamine into using it as the reward part of the guidance signal. This trick sends a learning signal to the rest of the brain suggesting that it learn to follow this guide, since the guides are assumed to lead to enhanced reproductive success. The listening neural structures basically don't question the wisdom of the dopamine fluctuations. The Heaven's Gate story exposes an "illusion" for this system, a limit in its reasonable operating range.

But where's the superpower in all this? Where is the "ideas veto instincts" bit? This sounds like the brain is constantly looking for ways to make ideas automatic—no matter what. But doesn't this remove our "agency," making what could be a series of pathological ideas into something automatic and unchecked—like an instinct? Like the "serve button"? Yes, it does. And if every idea, used as a guide to direct behavioral choices, became automated in this way, "we" would indeed disappear—sucked up into an automatic, unreflective set of computational mechanisms. Like a genie sucked back into its bottle. And while there would still be computations going on that cared about something (because of the valuation parts), there would be no agency living on top that cared much about anything. I suspect that there are pathologies, caused by disease or injury or maybe even environmental insult, that could create this condition. Damage to the prefrontal cortex correlates with a higher incidence of sociopathy— by diminishing the capacity of the prefrontal cortex to participate in the sustenance of new instinct-vetoing ideas that act to inhibit impulses. This latter explanation is not mine, but the computational mechanisms provide a quantitative setting to model and understand these processes in more detail.

Something like this happened with Heaven's Gate and other similar cults, and maybe all suicides—their entire cognitive world became an automated collection of complex mental habits. For whatever reason, there was no agency at the top questioning and intervening—screwing up the serve. Everything went on autopilot even though the cognition that remains is still complex. That's the distracting part. Just because a complex sequence of thoughts becomes automated does not mean that the sequence is simple; automatic does not necessitate simplistic. The whole process observed with cults used to be called "brainwashing," but now it's possible to apply a more detailed mechanistic story to their behavior, one that might suggest new behavioral or even pharmaceutical interventions.

We're not totally free. We start out being built by specific genes and then we accumulate experiential histories, all running on an efficient computational device that is constantly seeking to compress and automate our experiences, pushing them from the deliberative and explicit to the implicit. Especially the experiences that yielded important future returns—both good and bad. It's like we're fighting hoards of locusts—as individuals not too imposing, but collectively crunching away at our personhood. Tom Wolfe saw the locusts as the discoveries rolling out of neuroscience labs, and while we all understood his concern, I hope that I have deflected some of it by showing those aspects of valuation and choice that it overlooked. We have now seen that our soul, who we are, escapes neither the strictures of our genetic past nor our personal experiential past nor our shared cultural past. Our souls are not free-floating agencies with no rules or limits, but they aren't dead. They must choose in the face of our genetic heritage and our experiences. They must also cope with the strength and efficiency of the valuation mechanisms in our nervous system, especially since these mechanisms are always seeking to automate us, to take away a bit of what we call "us" and make it into a sleeker, more efficient part.

We are on a kind of personhood treadmill; we must constantly monitor ourselves, question our ideas, and probe our reasoning to test whether important parts of "us" have become automated, no longer susceptible to new ideas, new data, new events in the world. This notion of agency is interesting because we can now begin to probe it using imaging to expose its limits and those conditions when it is diminished. We are taking ourselves by the horns, staring straight at the agent (brain) that supports the minds we care about (us).

I BEGAN BY EXPOSING THE MYTH that imprecision, slowness, and noisiness are liabilities of brain function, "bugs" in computers. However, merely introducing the principles of efficient computation should do away with this myth. These are indeed the properties that any efficient computer should have—so now we can recategorize these "bugs" as evolved "features" of the need to compute efficiently. But since all computers run under resource constraints (all physical systems do), these same ideas apply to them as well. However, there is a subtlety here—simply understanding these features as adaptations to limited resources does not prescribe how to *distribute* resources among precision, speed, and signal-to-noise ratios (there are other related variables, but I'm sticking with the big three for clarity). The brain, and any other efficient computational device, should want to minimize energy dissipation *overall* rather than stubbornly insist on a fixed level of precision or speed. This is a distribution problem and efficiency is the product to be distributed. Of course, efficiency isn't "stuff," it's a measure of the way that limited stuff (resources) is doled out. Efficiency itself is just a computation of the way energy is handled, distributed, and stored. Every information-processing system needs a way to represent its efficiencies—in a biological cell, this might be encoded as the total amount of free energy available. But in other contexts, like an entire neural network, there may be variables that must be computed just to keep track of the distribution of resources. This issue has not been investigated in these terms in the nervous system.

The brain must possess dynamic distribution schemes if it is to be efficient overall. Some problems may require more speed and others more precision—and such needs will change with the problem at hand. Consequently, this resource allocation problem is not fixed once and for all, but changes with the situation at hand. Suppose that my nervous system committed "once and for all" to representing any number with something equivalent to three decimal places. Let's assume that I'm trying to throw a rock a great distance at a potential prey, and I really need some extra precision instead of more computing speed to keep my total energy consumption minimized. If my nervous system stubbornly stays with this fixed level of precision (three decimal places), my brain may not find the most efficient solution, because some process is insisting on the precision level. By limiting the precision, my brain may actually use more energy overall. So being inflexible about how much noise or how much precision or how slow to run a computation is not a good strategy. You have to figure it out as you go.

Dynamic resource allocation like this is the perceptual capacity we call *attention*. The idea of attention has been framed as a "limited resource"; however, the nature of the resource has not been connected directly to the problems of distributing energy across precision, speed, and signal-to-noise ratios. It has recently been pointed out that experiments in animals have not separated clearly the influence of attention and the influence of reward. While this complaint sounds like a technical detail for the experimentalists, I think that it is touching on a fundamental problem about energy distribution in the brain and the way that we now model the idea of reward and the idea of attention. Formerly, these two ideas referenced very different effects in the nervous system, but our perspective of efficient computation suggests a way to unite them. In our new lexicon, attention is not literally a resource but instead is the strategy (the algorithm) for distributing resources to computations. As a strategy, attention should express itself in many different forms, depending on the demands of the task. In fact, attention is equivalent to the concept of reward in many simple settings. This possibility is especially true in situations where creatures must decide how hard to work for particular options.

Yael Niv, a young computational neuroscientist, has addressed just this question using a class of reinforcement learning model. The question is "Given the current options in front of me, how hard should I be willing to work for each choice and how should this dynamically change?" Her approach is couched in equations, but that's close to the English translation. And of course, the distribution of vigor is exactly equivalent to the concept of attention in the behavioral experiments she has been modeling. Ironically, Niv's reinforcement learning model was crafted initially to ask how much effort a rat (those with fur, not suits) should put into alternative behavioral choices, but if the animal were carrying out a visual task, then her model also captures some properties of visual attention.

FINALLY, THERE IS THE CLAIM that computation, because it is merely the manipulation of strings of symbols, on its own carries no meaning. We touched on this assertion in the Introduction—second only to the problem of conscious awareness, it's the Portnoy's complaint of philosophers of mind. I've shown in various ways why value learning should emerge under the need to compute efficiently, especially when we recognize the need for goals in a highly variable world. These values get associated with symbols represented by your nervous system. We reviewed numerous experiments where a light or button press or sound or image, or even the abstract intention to trust, all accrued value because of their capacity to predict future reward. This fact means that these symbols (these representations), when placed in the context where they have reward-predicting value, carry a measure of their relative meaning to the organism. Now let's push the pedal to the floor.

The symbol represented in the brain is never alone; it's never one symbol followed by another and another, all meaninglessly careening from one system to another. That model of computation for our brain is wrong. It is not the style of computation to which the first biological nervous system committed. Instead, biology never separated the bare symbol from its overall value to the nervous system. Every symbol carries along its value. Instead of a stream of symbols alone, computation in the brain is a stream of symbol-value pairs: one computational object for the price of two. And the value part is not just an empty number, although it represents a number that embodies long-term judgment, that is, a representation of the long-term value to the organism of its sister symbol. This strategy gives meaning, valuation, to each symbol constructed and processed by the nervous system. Now compose a bunch of symbol-value pairs together in some kind of complex computation—each part of the computation will carry its value, and more importantly, the overall collection will also carry a value. So there are the semantics built into neural computation from the bottom up. This is at least one way to view the provenance of meaning—all the parts had meaning tags on them from the beginning, so building a composite machine or algorithm (remember the blurred distinction here from before) out of these parts built a structure replete with meaning. This is what fundamentally distinguishes biological computation from the meaningless symbol manipulation that we normally associate with the modern idea of computing. When we called the brain a value machine, it is "values all the way down."

And one of those machines chose this book.

ARE HUMANS COMPUTABLE?

THE STORY TOLD HERE leaves out many interesting areas of neuro-science. By omitting equivocations, I was able to bring into focus several principles that accounted for properties of neuronal firing and communication, algorithms used to seek and harvest rewards, and even algorithms for social exchange.

In this epilogue, I'm going to get slightly more technical and speculative. Some readers may choose to read on despite my warning, and others may choose to slide the book under the leg of a wobbly table. But both kinds of readers should have a better idea about why they can make that choice at all.

IF YOU USE A WORD PROCESSING PROGRAM, I can guess with near certainty that whatever it is, the program will run on many different kinds of computer chips. As a thought experiment, let's run it

on the "sweating chip." A quirky device with an extra physical property, it "sweats" when it accesses memory a lot. Like most computer
systems, an operating system is running on the "sweating chip" on top
of which the word processing program runs. When the chip has to access memory at a high rate, it begins to produce water—it "sweats." To
the manufacturer's great relief, the "sweat" causes no change in the
operation of the programs that it supports. The manufacturer doesn't
know exactly why this happens, but they have traced the "problem"
back to some contaminant in the air supply when the chip was made.
The sweating is incidental, but it's definitely an extra physical property of using this particular chip to run programs. If I monitored the
sweat production, I would have a crude measure of how hard the chip
worked while running its programs. Here's the punch line. There will
always be extra physical properties in any device that implements
computations, that is, a device that supports computations. If there
were not, then the device would have no pressing need for programs
at all—its physical states are exactly equivalent to programs.

An implementing device does not generally have "just enough"
physical properties for the computations it supports. The reason is
that situation would constitute a one-to-one relation between the
device's physical states and its overlying software. And while that's
theoretically possible, it makes any overlying software redundant.
The physical device is enough. The dynamics of its physical states
are in a one-to-one relationship with the overlying computations—
why not just let it exist and read out its physical states when an answer is needed? This might not be feasible in practice. It could be costly
for me to read out these physical states, and by mapping them one-
to-one to some kind of program, I have an easier time getting the answers I want—but this is a statement about my needs, not efficient

computations on the device. I bring this up to finesse a point here. When we say a device is implementing a computation, we must make a distinction between devices that run computations and have lots of extra physical properties left over (the vast majority of devices) and those that *are* the computations (one-to-one relation between computations and physical states). The translation of genes on DNA is a terrific example, because the power of DNA as a computer that implements life relies on the "extra physical properties" idea.

The genetic code is a relationship between triplets of base pairs in DNA and amino acids—the building blocks for proteins. When I learned about it in high school, I was amazed. The code was presented as a simple lookup table, just like one of those maps with letters marking rows and numbers marking columns. You could start with the amino acid and look in the table to find the three letters, the three bases that coded for it. Alternatively, you could start with the three letters and find the amino acid. Proteins built by chains of amino acids carry out every conceivable job there is to do in your bodies and brains. They are ionic pumps, struts, catalysts, protective coatings, ferryboats for molecules, rotors, timers, parts of informational networks, and on and on. The discovery of the genetic code, like uncovering the double-helix structure of DNA or sequencing of the human genome, marked an important moment in our history as a species. Organized dirt—us—now possesses the ability to probe the very information-bearing structures that organize us and our thoughts.

Here's the secret about the genetic code. There is no "fire" in it alone. It is an "amino-acid lookup table" that prescribes the order of amino acids to hook together into a chain. Remember that a lookup table is a computation, a patterned relationship between one set of symbols (the bases) and another set of symbols (the amino acids).

The remarkable functions of proteins are actually carried out by something else—the physical properties of amino acid side chains. Each amino acid has a unique fingerprint, an identity, defined by the physical and chemical properties of its side chain. These side chains form the basis of a way to group amino acids together—some act like bases, some act like acids, some like to dissolve in fat more than in water, and so on. Linking amino acids together builds all these remarkable protein machines only because the side chains are dragged into a really small space next to one another where they can interact. And this is important—if I placed a series of amino acids on a table-top one millimeter from each other, nothing would happen; no funky protein properties, no ionic pumps, no catalysts, and so on. However, if these amino acids are bonded together in sequence (by peptide bonds), the side chains get really close, and fireworks happen. A working polypeptide is made. I described how the order of amino acids, specified by the lookup table (the genetic code), was a computation. But what about the extra physical properties hauled in by the side chains? Are these properties all computable? There is a practical impediment to answering this question.

The practical issue is that the answer doesn't matter much because it's not worth it. The question as framed begs the question of costs. It would cost too much in space and time to describe all the possible physical interactions of the side chains, even if the description were written down using an "atom-sized font." DNA within a cell could not embody the description. Its own "multiatom-sized font" could not possibly specify the detailed physical interactions needed for any particular protein because the number of possible physical states among amino acid side chains is unimaginably vast. To compute them—to write them down using DNA base pairs (or any other physically viable symbols) would entail a description that would never fit in the tiny volume of an actual cell. And besides, even if we could write down these physical interactions using some kind of tiny symbols (maybe atomic states), the computation might not possess the "extra physical properties" required for the protein to function. Consequently, transliteration of an amino acid sequence into another lexicon would have to possess nearly the same set of "extra physical properties" in order to make working proteins. I suspect that evolution may have tried a range of lexicons, using different numbers or kinds of amino acids, but settled on what we find today.

So, instead of computing the physical interactions of the amino acid side chains, nature discovered an organizational scheme that specifies (computes) the order of amino acids, but relies on the fact that each amino acid also drags along extra physical properties. The computation exploits the extra physical properties that come along for the ride and this implicitly puts the properties to work. I suspect that there was not an option to compute the physical properties themselves because that computation would not fit on this planet.

The proteins' properties couldn't be computed (in the Turing sense) because there is not enough time or space available on this planet or perhaps even in our galaxy. What about an "in principle" argument? Could the physical properties and interactions of side chains that breathe life into proteins be computed by anything? I don't know, but my guess is no—I think that they are probably uncomputable or at least possess uncomputable capacities. This speculation, if true, suggests that our bodies have been computed by the algorithm of evolution, but that computation is best thought of as a way of organizing together a host of uncomputable properties to do complex tasks. The extra physical properties brought together by connecting amino acids together suggest that our bodies are replete with uncomputable parts. And our bodies include our brains, so the same conclusion would have to hold there as well.

I'm not suggesting something mysterious here. The DNA is a shorthand recipe for dragging all these complex physical properties into the same general vicinity, which allows them to interact. It prescribes only the order of the amino acids, not the physical interactions that will ensue once the amino acids are in place. The available alphabet for these computations is crucial; the primitive symbols must drag along the uncomputable objects, objects that could not be written down using any finite sequence of symbols. I know that amino acids can do this job, but I currently have no idea whether any other physical symbols possessing uncomputable properties could also do the job.

Uncomputable functions aren't so strange. Once Turing published his original paper, it was quickly understood that the vast majority of functions are uncomputable. The computable ones are the exceptions. My proposal here is that evolution bumped into this same feature of our world and discovered a computation that specifies which collection of uncomputable properties to bring together to perform some function: to build a protein.

This is where all the extra "computational" power in organisms originates—it comes from a specific pattern of computable and uncomputable parts. The computable strategies include information storage in DNA, the genetic code, and so on, which act as prescriptions for how to organize the uncomputable parts. Thus we are composed of two types of "patterns," one computable and the other uncomputable. A pattern of uncomputable parts sounds contradictory at first. Think of uncomputable beads on a string. The beads can't be computed, but these little uncomputable beads can be organized along a string and in a specific order. It's just that the inner workings of a specific bead can't be computed. And here the beads are the vast collection of physical possibilities possessed by amino acid side chains. So while humans are not computable, they use computations to organize their computable and uncomputable parts. I suspect that this will make simulating ourselves difficult, but not impossible.

This same general perspective on uncomputability has been arrived at by Roger Penrose and separately by David Deutsch, but starting from very different points of view. Penrose's ideas have been explored in a series of well-written books, but they generally lean on Gödel's Theorem. This theorem is an amazing result published in 1931 by Kurt Gödel four years before Turing's results on the halting problem. Gödel's results were also transcendent, but cast in a form initially less able to provide clear insight into the nature of our thoughts. The important outcome is that Penrose concludes that there must be uncomputable components contributing to the function of our brains and the minds they implement. Penrose's ideas go farther than just asserting the need for the uncomputable to account for the totality of our mental operations; he hypothesizes a physical basis for the uncomputable parts. Deutsch's story seeks to synthesize evolutionary theory, computational theory, and physical theory into one big explanation of why the universe has the form it does. In his book *The Fabric of Reality,* among other things he observes that our best physical theories are uncomputable and yet they describe physical processes that do occur. His conclusion, rightly or wrongly, is that the uncomputable parts are taking place in other universes. This conclusion depends on a description of reality called the multiverse model. And if true, it would suggest that our brains use more than our own universe to arrive at our thoughts—wild stuff, but possible. To my mind, the brilliance in their efforts is their stark reluctance to take anything for granted, so their minds lead them wherever they feel the facts warrant. They may turn out to be terribly wrong, but I think that their efforts reflect science at its best—taking nothing for granted.

In contrast to these efforts, my epilogue arrives at the same conclusion, directed primarily by some simple hints concerning the size of DNA, the mapping that is the genetic code, and the vast size of the functional space represented by the amino acid side chains that underwrite all the hard work carried out by protein. One hint is that even an atomic-sized description apparently lacks the "descriptive power" to write down a creature or even a cell within a reasonable amount of space. Also, let's not forget that the side chains are made of atoms and these have vast state spaces available to them as well. For example, large proteins tend to have lots of mobile electrons associated with them, and the state spaces available to them also can present a problem for Turing computable descriptions. Standing atop these data, I arrived at the speculation that we must be composed of uncomputable and computable parts. Nature can't compute an entire cell; however, Nature can certainly compose a cell by organizing uncomputable components together by means of a computation. Evolution found a mapping between a computation and uncomputable parts—humans aren't computable, but they can be organized by a computation.

This possibility means that we must deepen our quantitative understanding of what happens when we describe something—in words composed in paragraphs or in abstract symbols collected into equations. And it's here that I think that psychology has something to teach physics.

NEAR THE END OF his book *A Brief History of Time,* the physicist Stephen Hawking writes, "Even if there is only one possible unified theory, it is just a set of rules and equations. What is it that breathes fire into the equations and makes a universe for them to describe?" First, this is a magisterial thing to say, wondering what exactly science has wrought in its descriptions. I can't match profound statements with Stephen Hawking, and so I won't venture into this turf. But I will wonder over a more practical issue contained in his quote by ignoring the last part about the universe and making a simple analogy. Suppose I have a description of a nuclear power plant. I have never personally seen a full description of a nuclear power plant. Such a description would presumably occupy thousands of pages of printed text, equations, and diagrams. For our thought experiment, let's simply reduce a nuclear power plant to three main components: a reaction chamber containing fuel that reacts and heats up water surrounding it that flows around in a high-pressure waterline, a second low-pressure waterline that exchanges heat with the first line, turning its water into steam, and the use of the steam to turn the turbine of a generator that makes electricity.

Nuclear reaction → heat water under high pressure → heat water in another line to make steam → turn turbine to make electricity.

The idea is simple; actually making it work is hard.

Now ask a strange question. Why don't the equations describing this process get hot and the paper on which they are written burst into flames? They don't drag along enough extra physical properties to make this happen. They are merely a compressed, computational description of how to organize certain kinds of other parts (containing the extra physical properties) so that all the heat can result. OK, so you can't make a real nuclear power plant out of just the computations that represent the operation of a nuclear power plant. Like the amino acid story for proteins, a working plant requires that the computations be written with symbols that carry in other physical properties.

Imagine that we implement the pen-and-paper equations with scale-model parts. Suppose we build a tabletop-sized model of a nuclear power plant using a really small reaction chamber, tiny waterlines, and so on, but following very precisely the pen-and-paper description (the computations) to direct its assembly. But in the reaction chamber, instead of putting some nuclear fuel, we insert a slip of paper with the equations representing the nuclear reactor written on them. Will things get hot? No. There are no extra degrees of freedom in an equation on the written page, nothing leftover to wiggle around and produce heat. The equations won't get hot because they are expressed as symbols without any unaccounted-for degree of freedom. They won't get hot because our current arrangement has not brought into close proximity materials (symbols) possessing extra physical properties that interact together to produce extra wiggling around (heat). To make our scale model work, we must add to the reaction chamber something that produces heat in a way modeled by the equations.

MODERN PHYSICAL THEORIES are beautiful and practical. They are beautiful because of their mathematical structure and they are practical because they have extended our view of the natural world. They equip us with new ways of seeing the world around us. This is true whether or not you are trained as a physicist and possess structures in your brain that allow deep insight into the objects described by physical theories. As long as someone can distill the theories into a form consumable by the average human's psychology, the products of physics can be made available to us all. We have all been influenced by these descriptions whether or not we are explicitly aware of it. But suppose that physical theories were never digested in this way. Suppose they were never digested into a form that could be understood by anyone except a privileged few who shared a kind of secret language with one another. Or worse yet, let's get extreme and imagine the output of an important physical theory in a form understood by no human; too strange a description to run on the software of normal human psychology. These may have been discovered, but we won't hear about them. The reason is that the "output" of all physical theories must at some level be consumed by the psychology of humans. And we have seen throughout the book that this software constructs a world not for the purpose of consuming mathematically expressed physical theories, but for staying alive, recharging another day, and reproducing.

The way human psychological constructs have influenced the form of physical theories is an issue that has not been formalized in the context of understanding the theories. Take the way our mental constructs our perception of the temporal order of events. sion that one event follows another is at least a psychological construct, since even this very basic-sounding capacity is subject to illusions, quirks of its operating range.

In a clever series of experiments on time perception, David Eagleman has demonstrated this fact. His experiment is simple. While watching a computer screen, subjects are asked to press a button where a single flash of light at a time occurs near the button press. Sometimes the light comes after the press, sometimes before, and sometimes they occur at the same moment. Six seconds later subjects are asked (by the computer) to report which came first, the press or the flash. Simple enough. And important, as Eagleman is quick to point out: "Suppose that you are walking through the woods, you step and hear a twig crack. Which came first, the step or the crack?" The difference is one situation could translate into life or death and the other is the "normal" order of events that happens when one is walking through the woods. The normal order of events is step-crack, but if you detect crack first, a predator (or something scary) could be tracking you. The ability to detect and respond to the difference is essential for survival; however, this ability is not always reliable. Eagleman has discovered that if, while subjects play many rounds of this task, he secretly injects a delay between the button press and the flash, subjects will not notice the delay. The experiment hums along fine and the subjects' nervous system reencodes what before and after mean. Now when a press occurs followed by the light at a time shorter than the injected delay, then subjects perceive that the light came first—before the press. This report happens even though the "physical order" of events was the opposite.

Eagleman's experiments have lots of interesting twists, but the point for us is that time perception is a construct—like everything else in our mental operations. Does this mean that there is no "time" out there in the external physical world? This is not necessarily true, but it should call into serious question those parts of physical theories that we find "intuitive," since our "intuitive" has been constructed simply to keep us alive and reproducing and is not particularly well designed to consume mathematical theories of the physical world. But I think that Eagleman's demonstration, and the output of many other experiments on human perceptual capacities, should make us question why physical theories take the forms that they do. In order for the output of any physical theory to be "consumable" by human perception, its output must at some point be aligned with our capacity to understand it. A mathematically expressed physical theory has always had perceptual constraints built into it. Humans sit in front of the theories and make decisions with them and about them. For a theory to be useful to this consumer, even a highly trained consumer, parts of the theory must be a liaison back to our perception, while other parts must be describing something "out there." We're the consumers of any theory's implications and we have operating ranges in our ability to think about "before," "after," "spatially separated," and so on.

In conclusion, I'm not going to propose anything philosophical, rather something concrete and practical. We should now attempt to factor physical theories cleanly into two parts, the perception part and the physical part. In the early and middle parts of the twentieth century, there was not enough quantitative understanding of certain perceptual domains to get serious about factoring out the human perception part built deeply into physical theories. If the output of a theory must align with my capacities to comprehend it, and my capacities are engineered to navigate life on this planet, then it's reasonable to claim that my perceptual capacities have "clamped" and constrained the form of the theory. This "clamping" may propagate its influence through many aspects of the theory in ways difficult to extract later. So let's stop doing it implicitly and make an explicit attempt to extract our psychology from our physical descriptions. There is a natural and new kind of alliance brewing between psychology and physics.

Acknowledgments

I would like to acknowledge the National Institute on Drug Abuse, National Institute of Mental Health, National Institute of Neurological Disorders and Stroke, The Kane Family Foundation, and Baylor College of Medicine for support of my work over the years. In addition, the support of the Spencer Foundation and Deutschebank enabled my stay at the Institute for Advanced Study in Princeton, where I was able to finish the book in peace and among a new collection of colleagues. I would also like to acknowledge the pursuit of science. As a kid I did not really know science existed as a way to produce new knowledge and put food on the table; in many ways it rescued me and continues to do so today. I would like to acknowledge a collection of patient individuals who either read portions of the manuscript, discussed the ideas with me, or listened to me pitch some new way to present things: Ron Fisher, Peter Dayan, Steve Quartz, Brooks King-Casas, Ann Harvey, Damon Tomlin, Terry

Lohrenz, Telicia Montague, Amin Kayali, Linda Troiani, Jian Li, Donne Petito.

I would also like to acknowledge some important colleagues with whom I have interacted closely over the years and who have influenced either the content, tone, or ideas expressed here: Peter Dayan, Terry Sejnowski, Steve Quartz, Jim Patrick, Dan Johnston, John Dani, Antonio Damasio, David Eagleman, Patricia Churchland, Richard King, Sam McClure, Jon Cohen, Michael Friedlander, Francis Crick, Joe Gally, Gerald Edelman, and Nathaniel Daw.

Thanks also go to my newfound colleagues in economics with a special acknowledgment to Colin Camerer, Drazen Prelac, Dan Ariely, Kevin McCabe, and Ernst Fehr. I would also like to thank my dog, Scruffy, for not folding the paper into a piece of origami—that act would have ruined my entire story.

Lastly, I would like to thank my book agents, Katinka Matson and John Brockman, and especially my editor, Stephen Morrow, whose persistence forced me to grind away the "science-speak" from the book so that everyone could understand what I was trying to say. His efforts improved everything.

Endnotes

INTRODUCTION

x *but identifying evolutionary constraints is just one aspect of understanding how we think* Evolutionary arguments are indispensable in thinking about our thinking, especially in their pointing to a class of computation that needs to take place. For example, a need for computations that amount to "cheater detection" has been identified by evolutionary psychologists. However, identifying "cheater detection" as essential is just one step in understanding the neural basis of this computation. A complete understanding that connects to underlying neural computations requires that we translate these ideas in specific proximate (moment-by-moment) mechanisms. Over the last twenty-five years, a number of excellent books have emerged covering issues related to sociobiology, now called evolutionary psychology (Wilson, 1978; Barkow et al., 1992; Dennett, 1995, 2003; Pinker, 1997, 2002; Buss, 2004; also see Ridley, 1997, and Hauser, 2000). *Why Choose This Book?* aligns with the majority of claims in these books), but instead focuses on the proximate mechanisms underlying our psychology—those implementing neural mechanisms whose psychological function can be exposed through the use of computational models.

x *Theodosius Dobzhansky wrote, "Nothing in biology makes sense except in the light of evolution"* Dobzhansky, 1973. The same basic point made from two separate perspectives can be found in Williams, 1966, and Mayer, 2001.

x *address every part of our neural and mental function as information processing—computation* The mathematician John von Neumann was a central figure of the early and mid-twentieth-century efforts to mathematize all subjects, including those dealing with human thought and automating it. Von Neumann was simply amazing. Von Neumann and Paul Dirac each systematically mathematized quantum mechanics (Dirac, 1930; von Neumann, 1932). But von Neumann went on to found modern game theory (von Neumann and Morgenstern, 1944) and automata theory (von Neumann, 1966), and did foundational work in numerous areas of mathematics. Automata theory was the precursor to cellular automata and other computational approaches to life and thought. It does not generally include the capacity for quantum computation and is therefore likely to be incomplete (e.g., Wolfram, 2002), a position supported strongly by Deutsch, 1997. Von Neumann was as singular an intellect as Turing. An account of his life and exploits can be found in Macrae, 1992. From these early beginnings, work on machine intelligence developed mainly along the lines of engineering problems or what is now known as artificial intelligence.

For neuroscience, modern computational approaches did not emerge in full force until the 1970s and early 1980s (Marr, 1969, 1970, 1971, 1975; Marr and Poggio, 1976, 1977; Sejnowski, 1976ab, 1977ab; Hinton and Anderson, 1981; Hopfield, 1982, 1983; Koch and Poggio, 1982, 1983; Ballard et al., 1983; Hopfield and Tank, 1985, 1986; Ackley et al., 1985; Poggio et al., 1985; Hinton et al., 1986; Sejnowski et al., 1986). Lastly, I personally cannot leave out the PDP group who inspired neuroscientists, engineers, and others in the mid-1980s (Rumelhart and McClelland, 1986).

Collectively this work was not simply an effort to use computers to model brain function but instead to understand the brain as a special kind of computational device. The distinction is important—even today, especially as we start to consider seriously how the brain uses computable and uncomputable parts to carry out its functions. Modern efforts in computational neuroscience find many, if not most, of their roots in this early work and its relatives.

This early work was also propelled by the recognition in the philosophical community that ideas about the mind needed to be grounded in the actions of the brain, an effort led during this period by Patricia S. and Paul M. Churchland (P. S. Churchland, 1982, 1986, 1988; P. M. Churchland, 1981, 1984). Their efforts significantly broadened the scope of the issues addressed by computational neuroscience (see Churchland and Sejnowski, 1988, and Sejnowski et al., 1988).

CHAPTER 1: COMPUTERS THAT CARE

2   *Dönitz the warrior chose to kill, but he probably did care* Dönitz, 1997.
3   *And, at one time or another, computer programs have pushed us all to the edge* This sentiment is aptly expressed by the character Howard Beale (played by Peter Finch) in the Sidney Lumet–directed movie *Network* (1976). Beale, feeling that modern technology was dehumanizing the world, transformed from network

anchor to a nightly "ranter" with an evangelical mantra against the modern world:

> Beale: [shouting] You've got to say, "I'm a HUMAN BEING, God-dammit! My life has VALUE!" So I want you to get up now. I want all of you to get up out of your chairs. I want you to get up right now and go to the window. Open it, and stick your head out, and yell, "I'M AS MAD AS HELL, AND I'M NOT GOING TO TAKE THIS ANY-MORE!"

Many of us have reached a similar stage with our computers.

5   "... *If he is to be solely a Scientific Specialist, he is wasting his time at a Public School.*" Hodges, 1997.

7   "... *if a tape bearing suitable 'instructions' is inserted into it.*" Turing, 1936.

8   *your mind is equivalent to the information processing, the computations, supported by your brain* Computational theory of mind has been developing all through the twentieth century and space prevents anything remotely approaching a review. The idea developed in parallel in several disciplines—artificial intelligence, control theory, modern connectionist neural network approaches, and in the philosophical community (e.g., Putnam, 1961; Fodor, 1975).

8   *The philosophers are right about one thing, the meaning part is missing* Searle, 1980, 1984.

9   *until the last century, it has been difficult to question mind-stuff very clearly* Even the clearheaded William James had trouble defining the mind, although his use of language maintains a practical, empirical tone common in modern cognitive science. James, 1898.

10   *is incompatible with a literal mountain of facts about inheritance and the evolution of biological traits* Other than our current incapacity to explain conscious awareness and other aspects of our mental lives, there is no data to suggest that our minds did not evolve like everything else about our bodies. Lots of alternatives are being explored—it's a problem that interests a wide range of minds (e.g., Churchland, 1984; Churchland, 1986; Dennett, 1991; Chalmers, 1996; Shear, 1998; Koch, 2004).

10   *And while CTOM does not account for all our experiences* Chalmers, 1996.

11   *This is not a new idea; evolutionary biologists and some philosophers of mind have had this general perspective for many years.* Putnam, 1961; Williams, 1966; Fodor, 1975; Dawkins, 1976, 1982; Wilson, 1975; Churchland, 1984; Churchland, 1986; Dennett, 1991, 2003.

12   *This idea began first in the mid-twentieth century in a famous paper by Max Delbrück and Linus Pauling* Pauling and Delbrück, 1940.

14   *vitalism emerged in some form in almost every culture and usually in association with prescriptions for life and medical treatment* Reill, 2005.

16   *consider popular author Ray Kurzweil's prediction that machines will become sentient in the next thirty years or so* Kurzweil, 1999, 2005.

16   *Turing's original insight is as singular as Darwin's idea about natural selection,*
     and like all great ideas, its simplicity hides its depth Many authors have ex-
     pressed this exact sentiment, e.g., Wilson, 1978; Dawkins, 1986; Dennett, 1995;
     Pinker, 2002.

16   *It has been said about Darwin's theory of evolution that it's the ultimate*
     tautology—the survivors survive Williams, 1966; Mayer, 2002.

17   *Hunting and gathering is simply not very efficient* Some good modern starting
     points—Maisels, 1990, and Diamond, 1997. Both accounts show that the rise of
     agriculture has a complex history woven together with social structures and
     shared values. The shift to agrarian cultures of various sorts was a critical in-
     fluence.

17   *And when we look at the components of life, cells, they are literal wonders of effi-*
     cient energy-handling Nelson, 2003.

21   *colleagues to produce a series of electromechanical devices called bombes for com-*
     puting (breaking) the Enigma codes Turing and colleagues used a method for
     breaking the Enigma codes that strongly resembles what are now called Bayesian
     belief networks, a method of inference thought to be employed by the brain for
     decision-making. Gold and Shadlen, 2002, discuss this connection.

CHAPTER 2: THE BRAIN IS (ALMOST) PERFECT

24   *I'm not sure where it originated, but like the whisper game in kindergarten* The
     whisper game is also called the telephone game. It's a game that demonstrates
     how easily a message, passed along a line of communicators, can become cor-
     rupted beyond recognition. A line like "I'm playing a rubber piano at the club
     tonight" might pass along a line of "whisperers" and emerge as "Mel Brooks
     was here."
         I believe that similar "rumors" about neural computation emerged be-
     cause on the features of speed and accuracy neurons fall miserably short of
     what modern computers can do. However, that misses the point, since neurons
     run on batteries and must make good decisions about how to allocate their
     limited energy resources. Modern computers are grossly oversupplied with en-
     ergy and never had to contend with collecting their own lunch nor deal with
     decisions about distribution of limited stored energy.
         Modern computing is already running into thermodynamic limitations as
     the size of components in computer chips gets smaller and smaller. This
     shrinkage has already led to serious reevaluation of power distribution in com-
     puter chips. My sense is that simply focusing on the chips will not be enough.
     Software design (at all levels) will need to "be aware" of the demands they
     make on new power-efficient chip designs. This evolution should speed up sig-
     nificantly in coming years as the demand for small, wearable computing grows.
     See Kurzweill, 2005.

25   *the brain somehow (usually through the miracle of parallelism) produces mar-*
     velously subtle perceptions and behaviors that no man-made machine can now

rival The implication is that the slowness, imprecision, and noisiness of neurons is overcome by running computations in parallel, and miraculously so. However, this solution runs into terrible difficulties because of the serial needs of a single body. A single body must jump out of the way, or duck, or decide to spring up onto a branch. No matter how many parallel computations are thrown at these problems, the answers must be ready in time and presented to other neural processors responsible for initiating these very serial actions. Behavior forces the system to collect answers at a particular pace, and if too much parallelism is employed, then it becomes a great burden to compile the results of gazillions ("lots") of parallel computations. Hence, parallelism only goes so far.

At the beautiful Rockefeller University in New York in the late 1980s, I had access to a very fast parallel computer called N-Cube that possessed 1,024 processors. My programs would run with enormous speed until I needed to produce output on a display device. The bottleneck of having to render results in a comprehensible form to a computer screen was by far the slowest step. Miserably slow. "Writing to screen in human-readable form" became the Achilles' heel of the entire system, and we used to work very hard to avoid complicated output. In essence, all that parallel power was wasted because the results had to sit around waiting for the display processors to catch up. If the machine designers had to worry about the extra energy used by computing faster than output could be delivered, I strongly suspect that they would have made this step adaptive—that is, ask the processors to compute only as fast as output was being rendered. If the N-cube's "life" had depended on getting this energy matching correct, it would be dead now. Serialization imposes a severe limit on the utility of parallelism. And bodies impose serious serial constraints.

25   *Go ahead and check your monthly bill* Reliant Energy in Texas charged me 16.2 cents per kilowatt-hour during the month of March 2006.

27   *constructing systems that use biological components in conjunction with human-engineered components* The large, organized effort to develop working neural prostheses—devices that interface with our nervous systems in an effort to augment or supplant a missing or perturbed neural function—has produced, for example, the cochlear implant for the deaf and hearing impaired. Deep brain stimulation from electrodes implanted in specific brain regions has been used to mitigate the motor symptoms associated with Parkinson's disease. Many more possibilities are now at the research stage. The National Institutes of Health maintains a Web site for their neural prosthesis program: http://www.ninds.nih.gov/funding/research/npp/

27   *I have chosen to confine my discussion to how things are done now, how decisions are made moment to moment* Actually, I'm choosing an even tighter "confinement" than this line might imply. There are a host of problems in classical decision sciences that deal with moment-to-moment decision-making. I'm ignoring these and focusing only on the way that energy limitations have constrained the

character of decision mechanisms likely to be present in neural tissue and available for use in moment-to-moment decision-making and valuation.

29 *A computation is the manipulation of any set of symbols according to a set of rules* This is the classical model of computation equivalent to Turing's original proposal. There is a one-to-one relation of classical computation and classical physics; however, we now know that classical computation is a subset of quantum computation. The physicist David Deutsch gives a lucid introductory discussion of this fact. Throughout this book, I stick with the classical model of computation and focus primarily on how a consideration of costs recovers lots of properties observed in real neural systems and serves to connect neuroscience to important parts of psychology. However, the parts out of which the nervous system is made are themselves capable (in principle) of quantum computations. In the Epilogue, I consider briefly (and breezily) the sense in which our bodies and brains can be viewed as "computations using uncomputable parts" through a discussion of computing the structure of a protein.

All modern computational neuroscience currently employs solely the classical model of computation; however, these efforts will soon have to incorporate ways in which biology has taken advantage of quantum computations. Turing understood the limitations of his "computation" ideas early on and actually proposed, and studied briefly, a class of device called an oracle machine (O-machine), in which classical computations took advantage of "oracles" whose internal operations were not describable, but that could be queried for answers at moments defined by a particular classical computation. See Turing, 1939. Wouldn't we all love to have an oracle or two available on request?

29 *picked up by the ear, and changed into patterns of electrochemical activity in the brain* I have avoided detailed descriptions of electrochemical transmission of information in the nervous system. There is a vast amount known about the molecular parts and their interactions that underlie electrochemical transmission. See Kandel et al., 2000, for an introduction to this topic.

30 *it does not explain the total experience of poetry* Emotions and first-person conscious experience have so far escaped sensible computational descriptions of the kind that I am emphasizing. And the experience of poetry involves both. Workable computational descriptions of emotional responses are on the horizon, but no one has yet upgraded our first-person conscious experience to a problem that can be addressed by mathematizable models.

30 *encoded into mathematical descriptions, and capable of reproducing (predicting) the outcome of experiments* The notion of computational descriptions is implicit in all of physics. Sequences of symbols are written down to communicate to others the outcomes of experiments or the relationship of one physical phenomenon to another. The vexing part for the casual consumers of modern physical theories is the nature of the "objects" described by modern theories. What exactly do these theories describe? And more importantly, how do physical theories incorporate implicit models of human psychology? As psychology and cognitive science mature over the next decade, the increase in formal models

in these areas may permit a more formal integration of physical theories with cognitive theories. I touch upon one aspect of this issue in the Epilogue.

34 *The "drain slowly" principle shows that slowness in computation should be seen as an adaptation, not a liability to be surmounted* All things being equal, spiky demands are bad because they tend to waste energy, and "softer-gentler" transitions are good because they waste less energy (sounds a little like nervous-system politics). Consider the concrete example of the charging and discharging of a capacitor. To a first-order approximation, the charge Q on a capacitor is proportional to the voltage V, $Q \propto V$, so their rates of change are also proportional: $\frac{dQ}{dt} \propto \frac{dV}{dt}$. But the left-hand side is just the current, and power is proportional to the square of the current; therefore, power $\propto (\text{current})^2 = \left(\frac{dQ}{dt}\right)^2 \propto \left(\frac{dV}{dt}\right)^2$. The power dissipated depends on the square of the rate of change of voltage across the capacitor. This is why spiky rates of change from one voltage to another are bad; their effect on energy dissipation is squared. Given this argument, the obvious strategy is to keep the system warmed up and make smooth transitions when transitions are needed. If charging and discharging a capacitor is important for information processing, then the voltage changes should be as small and slow as possible consistent with the computations required.

This same logic applies to biochemical reaction schemes. To save energy, biochemical reactions in your body should be kept near equilibrium so that when they progress in one direction or another, it takes less energy to get them started. They should also possess smooth transitions from one reaction to another. But slow and smooth is not always possible or even desirable. If I sprint across a field to escape a predator, my return is my life. Sometimes deviations from slow and smooth are indeed warranted.

I'm discussing the thermodynamics of computation, and over the last thirty-five years ideas about the cost of computing have developed in the physics community and have been applied to the physiology of the nervous system. Until the early 1960s, it was thought that computing on any kind of classical computational device had a cost proportional to kTlog2 per bit. However, in the early 1960s, Rolf Landauer at IBM Research published a paper suggesting that only erasure is the energetically expensive step, so if a device was willing to keep information around, it could avoid the expense of erasing the bits formerly saved in memory (Landauer, 1961). This argument was clarified by Charles Bennett, who showed that almost all irreversible computations (the kind that dissipate energy) could be represented by reversible computations (albeit more complexly rendered than their irreversible equivalent [Bennett, 1973, 1979, 1982; also see Bennett and Landauer, 1985]). This same idea was discovered by Fredkin and Toffoli, 1982. A reversible computation is one that can be undone, that is, you can retrieve the inputs from the outputs. And if you're willing to compute slowly, you can almost totally avoid dissipating heat when carrying out a reversible or near-reversible computation. The importance for this book will arise again below, but it relates to basic information transmission in the nervous system at synaptic junctions—it's apparently irreversible, and in an efficient machine this fact does not make sense unless there's some really good payback somewhere.

In the physiology community, ideas about energy-conserving computations have been around for as long as scientists have tried to model neural function. These efforts typically cast the "energy conservation" piece in terms of optimality against some overall computation the model system is thought to perform (e.g. Hopfield, 1982; Bialek, 1987). In parallel to these efforts, physiologists began to take the cost of neural computations seriously. For example, the work of Simon Laughlin (Laughlin et al., 1998; Laughlin, 2001; Laughlin and Sejnowski, 2003; Faisal et al., 2005) and William Levy (Levy and Baxter, 1996, 2002; Sangrey et al., 2004) has provided new insight into the cost of producing and transmitting neural impulses—one of the main forms of computation in the brain.

But avoiding spiky transitions is not a precise enough prescription. Spiky compared to what? There must be some ongoing measure of the value—the payback—of pursuing some energetically costly computation. And this measure needs to be dynamic, constantly reassessing whether one computational strategy should be pursued over another. This is why online valuation is such a critical function for a creature that runs on batteries that it must itself recharge. This is why our bodies and our brains are full of valuation mechanisms.

The harvesting of an exhaustible resource—like draining an oil field (Devarajan and Fisher, 1981) is a useful analogy. Extracting the oil has a cost, and the net return for extraction is the value of the extracted oil minus the cost of extraction. The problem is that as extraction proceeds, the costs may rise and the value of the extracted resource could change as well (possibly diminish).

For example, suppose we were extracting oil with a rope and bucket from an old-fashioned water well containing oil at the bottom instead of water. The cost is the energy to drop the bucket down to the surface of the oil, fill it, and lift it hand over hand back to the top. As extraction proceeds, the level of the oil in the well drops and we have to drop and retrieve the bucket over a greater distance—the cost of extraction increases with each bucket and so the net return diminishes. Here's the tough part: Suppose I'm not completely sure about the depth of the oil. How long should I keep investing in dropping the bucket and extracting the oil? It depends on: (1) the cost of dropping the empty bucket and retrieving the filled bucket, (2) the value of each retrieved bucket of oil, (3) the rate at which the cost of retrieval rises (or not), and "the mother of all factors," (4) my prior estimate of how much oil is there. My confidence in my estimate is crucial. There are formal methods for handling these problems, but none of them can escape the uncertainty introduced by my initial lack of knowledge about the exact depth of the oil.

35   *There are two main cell types in the brain, neurons and glial cells* The important issue to remember about the specialized cells that compose the nervous system is that they evolved from cells. They evolved from building blocks that already

knew how to (1) react to changes in metabolic demand, (2) to repair or replace senescent parts, and (3) to make copies of themselves. Once evolution learned to make nervous systems with wires (axons and dendrites) and highly adaptable connections (synapses), these features inherited all the self-monitoring and self-repairing functions that all cells possess. So these wires were always self-healing and dynamic. With these features in mind, it's not surprising that ongoing structural rewiring is one way that the brain learns and stores information.

35 *to make the neural impulses travel faster, significantly more energy would have to be expended by the axon (the wire)* Sangrey et al., 2004, calculates the energy cost of making action potentials travel faster.

37 *flexibly direct the storage and recall of information in the brain* The standard tale of electrochemical transmission goes thus: Electrical impulses, defined as rapid deflections in voltage across a cell's membrane, travel like "snaps in a garden hose" along neural fibers called axons and dendrites. At the connections made by the axons (synapses), these "snaps" cause the release of neurotransmitter substances into the tiny gap (synaptic cleft) between the axon and the target neuron receiving the connection—a cleft that's about twenty billionths of a meter wide. These substances move by passive diffusion in this gap and bind to membrane-bound receptors, both on the target neuron and on the axon's own synapse. This binding initiates a variety of electrical and chemical signals in the target neuron. (See Kandel et al., 2000.) The "snaps" that move along dendrites also open ionic channels, change the state of membrane receptors, and may track to those regions of dendrites most recently active (Magee and Johnston, 1997, 2005; Johnston et al., 1999). At the level of action potentials ("snaps"), there is a tremendous amount of detailed information known, but many questions about the information these spikes carry remain wide open (Koch, 1999; Dayan and Abbott, 2001).

The adaptability of axons, dendrites, and synapses highlights the fact that almost every identified structural level in the brain can change as a function of experience. The style of information-processing presents some really daunting problems, especially in light of the theme of energy efficiency.

For example, the battery-draining ideas above highlight a big mystery. Above, I outlined a "high altitude" version of electrochemical transmission, where action potentials (snaps) travel to synapses and cause the release of neurotransmitter molecules, which diffuse across the tiny synaptic cleft. But wait— diffuse? The mathematically inclined reader will notice an apparent deviation here from all my rhetoric about efficient computing. No one knows why the premier communication junction in the nervous system (the synapse) has committed to such a goofy communication channel—a diffusive one! All neuroscientists think that important information is encoded in the pattern of action potentials traveling along axons. But at the synaptic junction, this important information is converted stochastically into a sudden "bump" in the density of random walking neurotransmitter molecules, which hop around independently. Now, it's true that a twenty-billionth-of-a-meter gap (synaptic cleft size) is really small and that diffusion is fast on that scale, but those observations cannot get around the fact that this style of communication at these junctions *erases information and generates heat!* This loss presumably happens at every one of the $\sim 10^{15}$ synapses in the brain each time they transmit, and no one knows why. This is weird and I don't know what it's all about. It's not sufficient to "excuse the biology" and say that evolution only had "sloppy" components with which to mold these junctions. My opinion is that evolution *picked* this style of communication junction for a reason, possibly to introduce a calculated amount of uncertainty at every synapse (e.g., Glimcher, 2005; also see Lopera et al., 2006).

37   *more computational machinery than we have skimmed here, it's just not very well understood yet* Biochemical cascades, both within cells (neurons and glia) and outside them, have been characterized using a wide variety of physical techniques. However, their exact computational capacities are not well understood. In the past, the spiking behavior of neurons has been the most conspicuous computational channel to analyze, in part because spikes are easy to measure— they were amenable to electrical recording techniques a hundred years ago. We now see that it's also conspicuous because spikes are so energetically expensive—they simply must be carrying vital information in an efficient machine. However, modern techniques are now showing the vastly complex computational capacities inside single neurons and we should expect to see major growth in our understanding of these intracellular computational capacities in coming years (Abbott and Regehr, 2004). It would not surprise me if neurons could use these internal biochemical networks to emulate (model) the computations of every other neuron to which they were connected, since it's here (inside the cell) where internal models of themselves could be compared to the simulated models of their communication partners.

38   *They are smaller and they are less likely to produce errors* All things being equal, more steps means more chances to make a mistake.

38   *Wires are a particularly costly form of communication channel and their total length must be minimized in an efficient device* There are several kilometers of "wires" in any cubic millimeter of cerebral cortex. The thinnest are around one tenth of a micron (millionth of a meter). If they were any smaller, thermodynamic fluctuations in ionic channels would cause them to fire off action potentials spontaneously, potentially corrupting the messages carried by the pattern of spikes they support (Faisel et al., 2005).

39   *Lines drawn on a piece of paper can be represented as set of "drawing instructions" in an appropriate programming language* Notice that for this sentence to be true, there must be a close alliance between the programming language and the kinds of "drawing instructions" that it expresses.

39   *a three-dimensional animal sculpture might be represented as an extremely compressed collection of one-dimensional instructions* This is, of course, the trick embodied by DNA.

40   *decompression is the act of modeling the source of the compressed instructions* MacKay (2003) has excellent discussions of different concepts of compression and decompression. Here, I emphasize the idea of modeling the source of a signal.

41 *the information must be held in some kind of temporary storage (buffer), and this has a cost* Neuroscientists are not sure about how information is stored in the nervous system. One leading idea is that the "strength" of synaptic connections holds information, and while perhaps this is true to some degree, I would be shocked if it were the only physical implementation of storage in the nervous system—it's not nearly enough storage for the computations that must be carried out. However, no matter what the physical implementations, there will be a cost to buffering. It's just not possible to estimate this cost yet.

42 *A consistent "minimal wiring" picture emerges, although every rule is better seen as a statistical trend and not a "law of connection-making"* The idea that a "minimal wiring principle" is at play in nervous systems is as old as modern efforts to understand the "meaning" of neuroanatomical structures at all levels in biological nervous systems. The great neuroanatomist Ramón y Cajal noted that neurons, their processes, and their overall layout are likely to be answers to limited space and time constraints imposed on them (Ramón y Cajal, 1995), a view shared by modern work on the evolution and layout of the cerebral cortex (Allman, 1999; Felleman and van Essen, 1991). The best modern mathematical expressions of the minimal wiring idea began with the work of Graham Mitchison (Mitchison, 1991). Numerous other efforts have followed and examined a number of implications of "optimal" minimal wiring ideas. These include merely saving on wire (e.g., Chklovskii et al., 2002), the spatial layout of ganglia in invertebrate nervous systems (Cherniak, 1992, 1994), the detailed structure of specific cortical maps (Van Essen, 1996, 1997; Goodhill et al., 1997; Goodhill and Sejnowski, 1997; Chklovskii and Koulakov, 2000; Koulakov et al., 2001), the overall arrangements of cerebral cortical maps (Young, 1992; Young et al., 1995; Scannel et al., 1993, 1995; Stevens, 2001; Chklovskii et al., 2002; Klyachko and Stevens, 2003; Chklovskii and Koulakov, 2004; Cherniak et al., 2004), and the ways in which rewiring represents the storage and processing of information (Chklovskii et al, 2004). Even short-range geometric structure has received similar considerations (Shepard et al., 2005).

These results have been augmented by efficiency analyses of overall neural network structure and function (Olshausen and Field, 1996, 2004, 2005; Simoncelli and Olshausen, 2001; Milo et al., 2002; Sporns et al., 2004, 2005; Sporns and Zwi, 2004). This work is now in its third generation of finding the "correct" mathematical descriptions of this important savings principle.

42 *messages encoded as neural impulse patterns are cloned at each branch point of the axon and move out the daughter branches* The molecular mechanisms that support action potential propagation ("snaps in the garden hose") have been characterized to the point that highly accurate accounts of action potential propagation can be given. These accounts are now typical parts of course work in the area where students run simulations (at varying levels of description) of action potential initiation, propagation, and perturbation by drugs (e.g., anesthetics). Web sites: http://snnap.uth.tmc.edu/, http://neuronsinaction.com/

45 *their pattern should look "nearly random" through time* This general conclusion is a well-known result in information theory, and various incarnations of it can

be found in general texts on the subject (Cover and Thomas, 1990; MacKay, 2003). The modern idea of information compression and the ideal way to exploit available bandwidth in a communication channel arose with the work of Claude Shannon (Shannon, 1948). Although there were precursors to Shannon's work, he was the first to recognize the importance of making probabilistic models of message sources in order to compress messages. In addition, he introduced the idea of entropy to information theory. This was a bold move at the time, since the "information is physical" movement was still in its infancy. Shannon proposed that information is the degree of surprise that a particular message brings to a receiver of that message. And he framed that information H in a fashion directly analogous to physical entropy familiar from physics:

$$H = -\sum_{i=1}^{n} p_i \log p_i$$

where the likelihood of each message is p_i among a collection of n total messages. In his original paper (Shannon, 1948) he recognized that the "meaning" of a message was not pertinent to the problem of efficiently representing a message. His own words in that paper convey this idea succinctly: "The semantic aspects of communication are irrelevant to the engineering problem." And the engineering problem was to find a way to quantify the efficiency of a communication channel (of any sort).

46   *but these bursts do not last long (around ten milliseconds) and rarely exceed rates of four hundred to six hundred impulses per second* Kandel et al., 2000; Bear et al., 1996; Johnston and Wu, 1994.

46   *something called a Poisson interval distribution, and this prescription for choosing the forty times is theoretically the most efficient use of the bandwidth* This statement captures the idea that the coding strategy in the nervous system should utilize the available bandwidth efficiently (MacKay, 2003). However, this statement cannot be strictly true in a living, real-world nervous system. Neurons in the cerebral cortex possess a property called the refractory period. It's just what it sounds like: a time just after the neuron produces a spike (action potential) when it's "refractory" to making another for a short time. In fact, there are several refractory mechanisms. Based on our discussion, we might suspect them to be a way to enforce "slow" encoding strategies used by neurons. Second, neurons are also known to burst—they emit clusters of spikes at a very fast rate. Despite these features, neuronal spike generation is notoriously noisy. Yet the exact physical sources of this noise are not well understood, since cortical neurons can be shown to be near-deterministic engines of spike production when electrical current is injected directly into them (Mainen and Sejnowski, 1995; Mainen et al., 1996). It's almost as if there is a randomization box in there somewhere.

47   *everywhere he looks he sees biological systems working near the limits imposed by physics* This claim includes detection of signals by membrane-bound receptor molecules up to the detection and encoding of signals in the patterns of spikes

emitted by a neuron (e.g. Bialek and Schweitzer, 1985; Bialek, 1987; Bialek and Zee, 1988; Bialek and Owen, 1990; Bialek et al., 1991; Bialek and Rieke, 1992; Bialek and Setayeshgar, 2005).

This work examines in a principled fashion the efficient detection and transmission of information throughout the nervous system. Collectively, it represents hard-core physical thinking and the application of formal methods from physics to detailed issues addressed by experimental neuroscience. This work has provided some excellent first-principles approaches to the "encoding problem."

The correct first principles are not quite as clear when it comes to the decoding problem, since decoding will in large part depend on the actual "goals" of the moment—something that in practice may be difficult to guess. See Schneidman et al., 2003, and Dayan and Abbott, 2001 (Chapter 4).

49 *We would encounter the same problems that faced the recent mobile probes sent to the Martian surface* Mishkin, 2003.

51 *in the context of goals and guidance signals, is a hot spot for theoretical work pursued under the name* reinforcement learning Kaelbling et al., 1996; Sutton and Barto, 1998; Dayan and Abbott, 2001. For the mathematically inclined reader, Daw, 2003, provides excellent coverage of applications of reinforcement learning approaches to neuroscience experiments.

52 *Let's start with stimulus-response learning in animals, the basic framework of behaviorism* A short and to-the-point account of the pros and cons of behaviorism by philosopher George Graham: http://plato.stanford.edu/entries/behaviorism/

53 *a broadcast with a delay of about fifteen to twenty minutes* At the speed of light $(3 \times 10^8$ meters per second), the exact delay depends on the relative position of Mars in its orbit. However, this is a substantial delay if we want to send "driving instructions" to the rover. Imagine how difficult it would be to steer a car if the delay between turning the steering wheel and the actual tire movement was fifteen minutes!

53 *A behaviorist account of this episode would not guess that all sorts of new internal instructions were being loaded and tested* The engineering team was constantly working and reworking software fixes for *Sojourner* (Mishkin, 2003). During these updates, the device would not move; there were no external signs that anything was happening, but of course an enormous amount was happening inside the device. The point here is that a behaviorist account would not try to model these internal operations, and this is one reason that it's an incomplete approach to understanding the origins of behavior.

53 *it's defined by important experiences like eating food, having sex, drinking water, running from predators, and so on* For concreteness, let's focus on goals related to rewards. From Montague et al., 2004: "We refer to the state engendered by a reward as a 'goal.' Goals can exist at numerous levels and direct behavior over many timescales. Goals for humans range from the most basic (for example, procuring something to eat in the next minute) to the most abstract and complex (such as planning a career). In reinforcement learning, it is assumed that

the fundamental goal of the agent (learner) is to learn to take actions that are most likely to lead to the greatest accrual of rewards in the future. This goal is achieved under the guidance of simple scalar quantities called reinforcement signals."

54 *Framed this way, goals and desires (guidance signals) become well-defined mathematical constructs* Sutton and Barto, 1998.

55 *equating values to learned "wants" shows that the ability to learn to want something is a symptom of an efficient computational device* Dayan and Abbott, 2001; Daw, 2003; Montague et al., 2004.

55 *that valuation mechanisms can be correlated directly with physical measurements* Montague et al., 2006.

CHAPTER 3: MY RABBIT KNOWS WHAT TO DO

59 *There is always an implied machine in the background when analyzing the function of an algorithm, and it's better to be explicit about its assumed capabilities* Cormen et al., 2001, provides a great introduction to what an algorithm is. It addresses many of the issues raised more breezily in this book, but not particularly in the context of brains. It's also a great recipe book for common algorithms we all take advantage of whether or not we're aware of it.

But for true aficionados, those who want to know where it all comes from and how to place algorithm analysis in context, there is only one real place to begin and that's Donald Knuth's triumphant series The Art of Computer Programming. These books were ranked by *American Scientist* magazine as one of the twelve best physical science monographs in the last century.

60 *idea to prove important properties of all algorithms no matter where they were running* Turing, 1936, 1939; Britton, 1992; Herken, 1988. In his papers, Turing showed that his universal Turing machine (left formally undefined here) could compute anything that any other Turing machine could compute. From this, he conjectured that his "dog" could do anything that any other "dog" could do. In "computing" words, he conjectured that his machine could compute anything that every other computing machine might compute when operating on finite data using a finite set of instructions (program). The latter is just a claim, but seems reasonable, and is called the Church-Turing Thesis. Of course a human brain is at least a "slow" Turing machine but may go beyond Turing by employing uncomputable functions in its operation (see Deutsch, 1998; Penrose, 2006; Epilogue). Other than Penrose (1989), no one has made a detailed suggestion about how the brain or body might harness uncomputable functions in the service of carrying out complex functions.

Our understanding of quantum computation has risen meteorically over the last twenty years and the construction of devices to carry out quantum computations has focused attention on the practical issue of insulating them from thermal fluctuations from the "environment." In my opinion, the idea that quantum computations might be exploited in the brain has generally been ignored because of the (real) difficulty of insulating a brain-based quantum computation from the thermal noise present at thirty-seven degrees (body temperature). If true, this means that the thermal jostling reduces all the computation in the brain to classical computations. Demonstrations have gone the other way, that is, it was long ago shown that quantum mechanical systems could demonstrate universal Turing computation (Benioff, 1980, 1982; Deutsch, 1985).

For a gripping treatment of Turing's ideas and life, read the biography by Hodges, 2000.

62   *a system that can generate one virtual machine after another chosen to solve some learning or categorization problem facing the system* It has long been known that the prefrontal cortex plays a role in working memory, task or context switching, and executive control related to both. I'm adding here the idea that instead of separating these functions, another way to think about prefrontal cortex function is through the virtual machine concept. The suggestion is that the prefrontal cortex is capable of cycling through entire virtual machines for solving particular problems; each machine would use its own working memory contrived for a task, would have some kind of executive control, and would be able to task-switch within the context of its operating range. The analogy would be like cycling through entire personalities. This is of course just a speculation on my part, but it fits some of the conflicting data from this area of the brain.

62   *a system that searches for* simulated machines *best suited for a particular learning problem* This remark follows on the heels of the previous note, but commits to a mechanism for selecting a good fit of virtual machine and cognitive problem at hand. The idea is that the prefrontal cortex can make use of the reward-harvesting machinery in the midbrain and striatum to forage through an abstract space of virtual machines for solving some cognitive goal.

65   *allows the overall machine to reduce its internal communications and hence operate efficiently* If module A is reciprocally connected to module B, then both modules will need to contain models of themselves and their partner. Exactly how this is done is not important. The implicit assumption here is that modeling is cheaper than communication.

66   *possess models of all their talking buddies, and their talking buddies are those subsystems to which they are connected* My guess about all this modeling is that it takes place inside neurons and glial cells using the vastly complex biochemical signaling networks in operation there. This is speculation on my part, but I'm convinced that extensive modeling must take place and that intracellular networks have been overlooked for this function. In the brain, this would imply that each neuron could run anywhere from ten thousand to a hundred million dynamic models internally.

66   *cost of writing and erasing using the real enzymes that read and write the ultimate memory tape—DNA* Bennett, 1973.

66   *memory-intensive solutions to modeling, and that it would only erase informa-*
      tion when absolutely necessary Landauer, 1961; Bennett, 1973, 1982.

67   *without any significant feedback, which in this case is too slow to be of much use*
      anyway Wolpert and Ghahramani, 2000.

67   *known to generate forward models, and I suspect that the list will grow to include*
      all significant brain systems in the future Wolpert and Ghahramani, 2000.

68   *"... so novel oral-facial sequences may also benefit (as might kicking and danc-*
      ing). ..." Calvin, 1993.

74   *and we can now identify a neural system involved in these essential functions* It's
      more accurate to say that every neural system that has been identified must be
      able to model the expected uncertainty for the tasks it performs. Sensing ex-
      pected and unexpected levels of uncertainty along different dimensions is
      something with which the brain has had to contend for eons.

77   *the willingness to win or lose at another's expense can be traced to shared genetic*
      futures For the original argument, see Hamilton, 1964a, b.

78   *Trust—a mechanism that induces someone to carry out a social exchange despite*
      the risk of loss Here trust is a proxy for something like a degree of confidence in
      the expected response of one's partner despite the obvious opportunity for de-
      fection.

80   *Something very important remained the same* I'm discussing the properties of
      our point of view as presented to us. I'm describing something we can all do
      and can agree upon (or not). There is an analogy here for perceptual judg-
      ments about colors. In one setup, subjects are given a circle with a line through
      a diameter separating the top and bottom halves. A "color" is put in the top half
      and subjects are given three knobs to adjust until their perception is that the
      color of the bottom half matches the color of the top. These knobs could con-
      trol the illumination in terms of red, green, and blue or in terms of other coor-
      dinates like hue, saturation, and value. Humans fall into a fairly tight range for
      any given color in terms of knob settings. So whatever is going on "in their
      minds' eye" is deemed to be comparable based on the general agreement of
      these external settings. For my claims about "point of view" and the invari-
      ances associated with it, a similar kind of measurement could be done. Pictures
      could be given to subjects and they could be asked, "If you were lying on the
      ledge, would you be able to reach the apple in the tree?" The experiment could
      parameterize the distance to the tree and hook this to a knob. People could ad-
      just the knob until they were willing to say yes with confidence. The knob set-
      tings should change systematically depending on the distance to the tree from
      the ledge.

81   *to probe visual imagery capacities similar to the thought experiments discussed*
      above These thought experiments are just knockoffs from the classic work of
      Roger Shepard and Steven Kosslyn over a number of decades. They relate the
      capacities directly to principles on which physical theories stand. My reading
      of Shepard's early work suggests that this same general motivation lay behind
      his desire to find the right kinds of representational spaces for cognition (e.g.,
      Kosslyn et al., 2003; Shepard, 1990, 2001).

81   *is part of the tool kit that makes up our capacity for doing intuitive physics, and it most likely is present in all mobile animals to some degree* McCloskey, 1983; McCloskey et al., 1983.

82   *objects are those things that are invariant through translations in time* Something that does not persist through time does not qualify as an object to our psychology.

82   *The "object" that is unchanging here is the "you," or rather your nervous system's representation of your first- and third-person point of view* I'm referring to what your first-person account can report. I'm not making any claims about conscious awareness. A zombie might well have these capacities (e.g., Koch and Crick, 2001; Koch, 2004).

83   *he mapped the problems onto simulation capacities present in the minds of a wide audience* In my opinion, this defines the word *intuitive*: It's relatively easy for a relatively large crowd to simulate.

84   *of linear momentum (invariance to space translations), and conservation of angular momentum (invariance to rotations in space)* This puts into words the theorem discovered by Emmy Noether, which relates symmetries to conserved quantities. The English translation of her 1918 paper is by Tavel, 1971.

85   *The idea has been floating around for years that the visual system uses coordinate-free representations of visual objects* If brains of conspecifics use coordinate-free representations, then social communication over shared experiences becomes easier. See Shepard, 1990.

86   *that one should let all the plausible counterfactual simulations "bump into one another" to produce the best average outcome* Deutsch (1999) has pointed out that the decision-making apparatus attached to nonrelativistic quantum mechanics is consistent with the "existence of rational decision-makers." The value function (or preference function) in this case is the expectation value of an observable (Hermitian operators). I'm not sure what he meant by this precisely, but it raises an interesting point unrelated to physics.

In quantum mechanics (the kind we learn in college, not the really fancy stuff), a fundamentally weird thing occurs: Events that might happen (just like counterfactuals) actually influence the answers that do happen (answers recorded in an experiment). I've never really understood why things should be this way, but the mathematical game amounts to modeling underlying states as complex numbers, adding them up and letting them interfere as complex numbers do (interference), and then using a prescription (squaring their modulus) to obtain predictions about the relative outcomes to expect from an experiment. His suggestion in that paper was centered on another issue, but I think that he may have made a fascinating new suggestion for how to model decision problems. Minimally, we should reconsider why we use real numbers to model underlying decision variables.

CHAPTER 4: SHARKS DON'T GO ON HUNGER STRIKES

88   *going to the "next level"—to veto their powerful instincts to survive* Perkins, 1997.

89   *trading one life now for a new life (with related genes) in the next generation* See Hölldobler and Wilson, 1995, for an illustration of these points as applied to ants.

90   *and even other cell types in the nervous system adjust—all under the influence of the guidance signals* The classes of guidance signals that I have in mind are called neuromodulatory systems. In the mammal, they possess long axons, which deliver their neurotransmitters to widespread targets. These are clearly broadcast systems, computing some quantity, encoding it in their pattern of spikes (impulses), and broadcasting this pattern widely throughout the brain. These systems use neurotransmitters like dopamine, serotonin, acetylcholine, norepinephrine, and others. In the past, these systems were thought to modulate only slow events in the brain—"changing brain states." Although this is true, these systems all carry information at much faster timescales to guide learning, decision-making, and neural responses to uncertainty.

91   *Unlike those in your personal computer, software updates in the brain change neural structure* Hyman and Malenka, 2001; Hyman, Nestler, Malenka, 2006; and Kalivas and Vokow, 2005, review a host of molecular, cellular, and systemwide changes that occur during addiction or under the direction of neural signals whose actions are perturbed during addiction.

91   *There is a growing effort for computer scientists and neuroscientists to weigh in together on just these issues* Daw, 2003; Redish, 2004; Montague et al., 2004. These sources treat addiction formally as a problem in valuation and both employ reinforcement learning models.

91   *These four basic steps, when translated into their mathematical equivalents, form the basis of an approach to goal-seeking, learning, and decision-making called* reinforcement learning Sutton and Barto, 1998.

91   *It's a likely growth area between basic science and clinical science in the coming years* Egelman et al., 1998, has addressed Parkinson's disease in terms of a simple reinforcement learning model. Montague et al., 2004, introduces the idea of valuation diseases, that is, diseases that perturb underlying value functions. Redish, 2004, has specifically addressed addiction as a disease that perturbs underlying value functions. Daw, 2003, has introduced a number of novel ideas about reinforcement learning and disease in his Ph.D. thesis—a computer science thesis no less. Montague et al., 2006, has described in detail how reinforcement learning models predict the existence of fictive learning signals whose actions may well be perturbed by drugs of abuse—perhaps contributing to the feeling of "almostness" felt by many addictive gamblers, the feeling that one almost won the game. Yu and Dayan, 2005, have produced models of how neuromodulatory systems should handle uncertainty in the world. These are the same systems affected by cocaine and methamphetamine. So there is a growing connection between theoretical and computer-science approaches and psychiatric illness. I suspect that this trend will continue.

92 *These signals have been used in engineered systems and computer programs to equip these systems with goals, and to use the goals to guide learning* There is a succinct and excellent history of the area in Sutton and Barto, 1998.

92 *Ultimately, the goal of such systems is to produce autonomous, self-programming systems that achieve their goals flexibly and creatively* Moravec, 1990, 1998.

92 *fiendishly slow, and sometimes the system cannot solve even simple problems* Bertsekis and Tsitsiklis, 1996, and Sutton and Barto, 1998, contain examples that make this point from a variety of starting assumptions.

94 *TD-Gammon's exploits are fun to read about* See Tesauro, 1989, 1995; Tesauro and Sejnowski, 1995. Sutton and Barto, 1998, devote a section in their book to a discussion of TD-Gammon. The most amazing part about TD-Gammon was that it used a very basic version of temporal difference learning (Sutton, 1988), suggesting that backgammon is maybe not so hard rather than speaking to the craftiness of the learning procedure. Also see Barto et al., 1983.

95 *it's made of thousands of subparts, some focused on very specialized parts of chess games* Hsu, 2002.

97 *Sutton from psychology and Barto from computer science, but they were drawn together by the idea of systems that learn from experience* Sutton and Barto, 1998.

97 *Bertsekas and Tsitsiklis of MIT from the engineering side* Bertsekas and Tsitsiklis, 1996. Also see Bertsekas, 2003, for a modern pedagogical treatment of optimization problems.

98 *does not have to start out maximally ignorant, no "blank slate," and in these cases reinforcement learning systems can be made quick and efficient* No biological creature starts as a blank slate, a point well defended by Pinker, 2002. Recent examples of how reinforcement learning can be fast if given a good head start is Tedrake, 2004; Tedrake et al., 2004; and Collins et al., 2005. Schaal and Schweighofer, 2005, discuss the issues involved with a slant toward the neurobiology.

98 *These ideas have been analyzed and discussed elsewhere in excellent books* Sutton and Barto, 1998, contains many references into the relevant literature.

98 *This holds for rats, dolphins, worms, humans, bacteria, and so on* Pinker, 2002; Dennett, 1995.

99 *His book on the subject makes fascinating reading* Berns, 2005.

99 *The basic idea is that our internal goals get compared to real or imagined (simulated) experience to produce ongoing criticism in the form of identifiable critic signals* The critics could take numerous forms. The simplest looks like:

$$\text{Critic} = \text{current reward} + \gamma\,\text{next prediction} - \text{current prediction}$$

The critic signal is an ongoing comparison of the current prediction and the sum of the current reward and some fraction γ of the next prediction. This means that the system must be able to simulate out a bit to get the next prediction. This signal is "signed." When it's positive, things are "better than expected." When it's negative, things are "worse than expected." When it's near zero, things are "just as expected." Moment-to-moment changes in dopamine delivery, as measured in experiments, are consistent with dopamine playing this "critic" role (Daw, 2003; Montague et al., 2004; Wightman, 2006). The predictions in the pseudoequation above depend on internal valuation functions in an animal's brain.

100 *are now directing experiments in humans that seek to understand the scope of the critic's influence* The reward-prediction error hypothesis (Quartz et al., 1992; Montague et al., 1996; Schultz et al., 1997) changed the interpretation of transients in dopamine delivery. Formerly, dopamine release in the brain was equated with "pleasure" or "reward." However, dopamine release exhibits complex dynamics that depend critically on the situation and change dramatically with learning. The "reward" = dopamine idea, at least for dopamine transients, can't be correct. Instead, reward information (current reward) is combined with a change in value (next prediction − current prediction) to produce a composite signal that depends on reward information but is not equal to it. This same idea handles the concept of saliency, that is, when to assign more importance to a stimulus or internal event—and the obvious time to do that is when the prediction error is large. Large prediction errors are by construction salient and prediction errors near zero are by construction not salient.

102 *to produce a kind of* smart error signal—*driven by the present, informed by the past, and guided by the likely future*

$$Critic = (current\ reward) + (\gamma next\ prediction - current\ prediction)$$

could be rewritten as:

$$Critic = immediate\ feedback + (likely\ next\ value - current\ predicted\ value)$$

This critic is used to learn the values listed in this expression so that they come to reflect the long-term experience of the animal as it tries to harvest some reward—any reward. Here I have labeled the reward in the second case as immediate feedback.

102 *Just like the modems, the brain takes advantage of unused bandwidth by packing multiple signals onto the same wires* Dopamine fibers are likely to carry many signals, not all of which have been identified yet. There are indeed anomalous responses in dopamine neurons not explained by the reward prediction error hypothesis; however, very little about the rapid dynamics of dopamine delivery is explained by the idea that reward = dopamine. See Daw, 2003.

103 *(2) a stored value function that represents a judgment about the long-term value of each state, and (3) a policy that maps the agent's states to its actions* Daw, 2003; Montague et al., 2004.

105 *Neuromodulatory systems indeed play a volume-control role, but they also direct ongoing valuation of experiences, decision-making, and even changes in working memory* Usher et al., 1999; Aston-Jones and Cohen, 2005.

106 *as one example of how the human brain integrates emotional states with rational thought, a key feature of being human* Schultz et al., 1997; Schultz and Dickinson, 2000; Montague et al., 2004.

106 *These neurons are emitting information all the time, even when they don't change their baseline impulse rate* Quartz et al., 1992; Montague et al., 1996; Schultz et al., 1997. Also see Friston et al., 1994.

108 *when he recognized a striking resemblance between dopamine neuron activity and error signals used in abstract reinforcement learning algorithms* I remember Peter Dayan tossing the papers in front of me in the break room, asking, "Don't you think that looks like a prediction error signal?" and a rapid discussion ensued. The agreement between Schultz's data and all the first-order theoretical predictions was remarkable. And it also explained why the experimental data was so confusing-looking. However, even the times when the neurons didn't change their firing made sense under a reinforcement learning perspective.

Now the "not funny" moment. It took us four years and about ten complete rewrites to get the paper published. Ouch. See Quartz et al., 1992, and Montague et al., 1996.

109 *The model showed that Schultz had discovered one of the central critic systems in the mammalian brain, and one that encoded its criticism in the delivery of dopamine* Schultz's discovery has opened the door to an enormous number of new experiments in humans and permitted the development of computational models of one critical function implemented by this important collection of neurons.

110 *thus opening the door to a variety of new theoretical analyses of brain function* Hanna Bayer and Paul Glimcher's work in this area has been essential for the field for two reasons. First, another lab has now independently confirmed Schultz's findings. Second, Bayer and Glimcher submitted the electrophysiological to quantitative analysis to address the question of whether dopamine neurons are emitting quantitatively prescribed reward prediction errors in changes in their firing rate or encoded in the duration of pause and burst responses. Bayer and Glimcher, 2004.

111 *it would adjust itself to learn about a new source of food, water, or sex* Some of the downstream structures that receive the signal have no reason to doubt whether it's directing them to a stash of meat or to the idea of a spaceship hiding behind a comet's tail. So they adjust themselves under the guidance of this signal without much questioning.

113 *Wolfram Schultz. The reward-prediction error signal (critic) was first identified using experiments in monkeys, where lights predicted the future arrival of juice squirts a short time later* For examples see Romo and Schultz, 1990; Ljungberg et al., 1992; Schultz et al., 1993; Hollerman and Schultz, 1998; and Schultz and Dickinson, 2000.

115 *"Should I run from the predator that I think I hear in the distance or should I continue eating away at my latest kill?"* See Montague and Berns, 2002. There are two points here: (1) These value proxy systems must put very diverse stimuli on common scales; that is, they must have some kind of common

internal currency for comparing eating food to running from a predator. So while such computations may be complex, at some point they must be put into the same coordinates. (2) Being able to assign values across a set of possible actions becomes a creature's proxy for the current "meaning" of these actions to it. In our efficiency parlance, this means that the value must measure the cost of the action and the return expected consequence on taking the action.

116   *We will answer this question lightly* I'm using this simply as an example to emphasize that nervous systems were bound to discover external currencies in some manner reflecting what they already knew how to do internally.

117   *But the brain must be able to decide which models are more valuable to it* So the game is to run simulations of the near future, value them against one another, and arrange a kind of prioritization of which trajectories are more likely to return the best yields—all computed along some kind of common scale. These simulations could in principle include complex sequences of actions. In some contexts, the simulated trajectories of experience may be more like iconic sequencing from one outcome to another, measuring the value of a sequence of outcomes without computing through all the detailed steps connecting the outcomes.

CHAPTER 5: THE VALUE MACHINE

120   *If* E. coli *were much smaller, they wouldn't even need to move to acquire food because it wouldn't be worth the effort* Bacteria live in a world where they are moving around in stuff that's about like molasses would be to us. Technically, they live in a "low Reynold's number" environment. The Reynold's number is a dimensionless number that characterizes the ratio of internal forces (the "$f = ma$" kind) to viscous forces (the frictional kind that increase with velocity). If bacteria were a lot smaller, they would profit just as well by simply "floating" around and banging into nutrients rather than investing energy into propulsive movement and decision-making (which some of them do). For the full story on bacterial motility, Howard Berg is the source to consult. Berg, 2004, provides an excellent introductory account with pointers into the literature.

121   *but our purpose here is to emphasize the* valuation function *built into this scheme* Berg, 2004. There is a clear connection between environmental conditions and the tuning of bacterial decision-making to match internal needs to environmental demands. Economics started early.

122   *The bacterium has a value function built into the way it handles its aspartate encounters, and based on this valuation, it makes decisions about running and tumbling* Bacterial sensing goes beyond what is required to merely internalize some nutrient like aspartate. Instead, extra steps and extra "handling time" are invested by the organism that aid it in modeling its local environment and its recent experience. This is what I term the "modeling" stream in the text—that part of aspartate-handling used solely to improve an internal model of the environment—and it's this processing that was the early form of "reward" sig-

nals. So for a motile bacterium, a reward signal is a special stream of information that costs energy to extract, but which yields better future returns for the organism. This is why the concept of a reward signal is more basic than the common psychological term—even successful bacteria committed time and energy to extract and utilize these valuable streams of information.

123 *I'm not so sure about feelings, and indeed much has been written on this subject* Damasio, 1994, 1999; LeDoux, 1998; Koch, 2004.

125 *They call their idea the dopamine gating hypothesis* O'Reilly et al., 1999, 2002.

125 *Research over the last fifty years has contributed to our dramatically improved current picture of prefrontal cortex function* Stuss and Knight, 2002. Also see Roberts et al., 1998; Otani, 2004; and Fuster, 1980.

125 *nineteenth-century railway supervisor whose ventral prefrontal cortex was obliterated in an accident* Damasio, 1994.

126 *a fact highlighted by neurologists Donald Stuss and Robert Knight in a recent book* Stuss and Knight, 2002.

126 *functions of the prefrontal cortex, and remarkably, reward prediction plays a central role in their story* Miller and Cohen, 2001.

127 *a theme known in the neural network community as a recurrent network* Brody et al., 2003; Gisiger et al., 2005; Wong et al., 2006; Miller et al., 2006; Eliasmith, 2005.

128 *like a Star Wars defensive screen for incoming information* This is basically a description of the dopamine gating hypothesis. See O'Reilly et al., 1999, 2002.

130 *came to the exact same conclusions on physiological grounds* Durstewitz et al., 2000.

131 *There is now mounting evidence that the dopamine signal is indeed used in this dual capacity in the mammalian brain* Montague et al., 1996; Schultz et al., 1997; Montague et al., 2004; Montague et al., 2006.

131 *Earl Miller and Jon Cohen illustrate the moral of this story with a revealing example* Miller and Cohen, 2001.

132 *O'Reilly, Cohen, and Braver propose a straightforward but powerful analogy with the reward-prediction error model* O'Reilly et al., 1999, 2002.

134 *There is also evidence that the striatum is involved in goal selection and maintenance* Graybiel, 2005.

135 *one central concept in the way we think genes change through time* Kimura, 1983.

137 *They engender feelings typically associated with rewards—a kind of satisfying, mild, yet fleeting euphoria* Damasio, 1999.

139 *The capacity for an idea to "plug back in" as a reward signal is such a powerful capacity that it must be subjected to numerous levels of control* The system must have checks and balances to prevent "any old idea" from acting with the behavioral power of sex or food. There is a fine balance between pursuing a novel idea about anything and pathological obsession.

140 *obsessive compulsive disorder, body dysmorphic disorder, or any other form of obsessive thinking* Our nervous system's guidance signals (like dopamine) form just one part of the systems involved in obsessive compulsive disorder and body dysmorphic disorder. These are complex problems involving multiple

neural systems; however, I'm pointing out here that for thoughts of any kind to gain behavioral potency they will most likely utilize neuromodulatory systems.

141 *neurons, synapses, membrane receptors, intracellular biochemical networks, and gene expression* Overview of the area: Hyman and Malenka, 2001; Hyman, Nestler, and Malenka, 2006; Kalivas and Vokow, 2005.

142 *This view is supported by recent experiments in rats that also impinge on this model-based approach* Phillips et al., 2003.

143 *while we were working at the Computational Neurobiology Lab (CNL) of Terry Sejnowski* http://www.cnl.salk.edu/

147 *they represent, in a single number, the long-term rewards expected from that state into the distant future* For the mathematically inclined reader the reinforcement learning models, as supported by biological data, take the form of simple value-learning models. Paraphrased from Montague et al., 2004:

> In the simplest TD models of dopamine systems, the reward prediction error depends on a value function that equates the value V of the current state s at time t with the average sum of future rewards received up until the end of a learning trial.

> $$V(s_t) = average\ total\ reward\ delivered\ from\ state\ s_t\ until\ the\ end\ of\ a$$
> $$learning\ trial = average\ [r_t + r_{t+1} + r_{t+2} + \ldots + r\ (trial's\ end)]$$

> There are two sources of randomness over which the above averaging occurs. First, the rewards in a trial $[r_t + r_{t+1} + r_{t+2} + \ldots + r$ (trial's end)] are random variables indexed by the time t, which for convenience might be thought of as "time," but is more accurately depicted as a parameter that keeps track of the order of experienced states (like a perceptual "ticker"). For example, r_{t+2} is a sample of the distribution of rewards received two time steps into the trial. The idea is that the animal can learn the average value of these rewards by repeating learning trials, and by revisiting state s_t sufficiently frequently for its nervous system to be able to estimate the average value of each of the rewards received from state s_t until the end of the trial. The second source of randomness is the probabilistic transition from one state at time t to a succeeding state s_{t+1} at a later time $t+1$. The value function, stored within the nervous system of the creature, provides an assessment of the likely future rewards for each state of the creature; that is, the value must somehow be associated with the state.

> However, it would be virtually impossible to make good estimates of the ideal $V(s_t)$ as it is now defined. This is because the creature would have to wait until all rewards were received within a trial before deciding on the value of its state at the beginning of the trial. By that time, it is too late for such a computation to be useful. This

problem becomes worse in real-world settings. Fortunately, the equation above provides a way out of this dilemma because it obeys a recursion relation through time:

$$V(s_t) = E[r_t] + V(s_{t+1})$$

This recursion relation shows that information about the value of a state s_t is available using only the value $V(s_t)$ of the current state s_t and the value of its successor state s_{t+1}. Until this point, we have been discussing the ideal case for V. However, as indicated above, V cannot be known exactly in the real world. Instead, an estimate V^* of V must be formed within the nervous system. The TD algorithm learns an approximation V^* of the value function V by employing a natural error function to criticize current approximation of the value function: the difference between the two sides of the above recursive equation (Bellman equation). This difference is called the prediction error or temporal-difference (TD) error.

$$\delta(t) = prediction\ error\ (t) = E[r_t] + V^* (s_{t+1}) - V^* (s_t)$$
$$\sim current\ reward + next\ prediction + current\ prediction$$

This TD error signal reproduces the phasic burst-and-pause responses measured in dopamine neurons recorded in alert monkeys during learning tasks. The next value of each adaptable weight $w(t+1)$ used to estimate V is incremented or decremented in proportion to the product of the current prediction error $\delta(t)$ and the current representation $T(t)$ of the stimulus responsible for the prediction.

$$w(t+1) = w(t) + \lambda T(t)\delta(t)$$

Here, λ is called a learning rate.

149 *This same TDRL model of learned value has been an influential psychological model of "wanting" proposed by Terry Robinson and Kent Berridge* McClure et al., 2003. This paper shows how "wanting" something becomes a learnable parameter whose value could be learned under the guidance of dopamine signals. Note that one of the equations in this paper has a typo.

150 *So the wants aren't measured directly by the dopamine signal, but differences across wants are* McClure et al., 2003.

150 *the model can address something as practical as voucher schemes to keep addicts from taking drugs* Redish, 2004.

150 *fast changes in dopamine delivery to different parts of the brains of rodents while they are freely moving about* Wightman, 2006. The work of Wightman and colleagues has revolutionized our ability to understand fast fluctuations in dopamine levels in the brain. This work has exposed an entirely new

timescale in which dopamine functions as a relatively fast communication channel. Not only has this group uncovered fast timescales of dopamine function, they are also able to make these measurements in freely moving animals—a maneuver that has opened up all sorts of new scientific questions related to reward-dependent action choice and addiction.

151   *that this new quantitative understanding and its descendents might translate into principled therapeutic interventions in addicts* Wightman's work single-handedly made possible the construction of detailed theoretical descriptions of how changes in dopamine neuron spike activity translate into actual dopamine release on fast timescales—Montague et al., 2004a.

154   *It's very much like Sigmund Freud's view of the paranoid personality—only the premise is wrong* The claim here is that the decision mechanisms are intact, that is, they respond in a correct way to the underlying value function. It's the valuation part that is perturbed, just as in the case of the TD model of addiction explored by Redish. Some feature of a situation may get excess value— like the concern over disease-causing germs. If the relative value of this possibility were way off, then constant hand-washing might well be exactly the rational solution. This would of course represent just one part of the problem, but it does offer a new perspective.

155   *The striatum is now known to be involved more intimately in selecting and pursuing goals, and in the processing of reward-related information* Graybiel, 2005, gives an overview of our rapidly changing views for this area of the brain.

160   *well known for his pioneering work on decision-making with longtime collaborator Nobel Laureate Daniel Kahneman* Tversky and Kahneman, 1974, 1981; Kahneman and Tversky, 1979.

160   *Tversky and his colleagues showed that neither the length nor the frequency of shooting streaks exceeded that expected at random* See Gilovich et al., 1985. Belief in the "hot hand" is strong, and so this work created a flurry of public and academic interest.

CHAPTER 6: THE FEELINGS WE REALLY TREASURE

164   *an essential capacity for any flexible, social species, and a capacity known to break in humans* Weigelt and Camerer, 1988, and Berg et al., 1995, proposed two-person economic games that probe the essentials of trust. Using a modified version of the Berg et al. trust game, King-Casas et al., 2005, have carried out a large-scale functional MRI study of trust, but adding the twist of recording activity in both interacting brains during the game (http://www .hnl.bcm.tmc.edu/trust.html). Also see Delgado et al., 2005.

165   *Ultimate goals are something that the economist Adam Smith would recognize: They are the invisible hand of fitness, and they don't really care about individuals* Ultimate goals here would refer to reproductive fitness, something that ultimately guides selection, but is not necessarily "felt" moment by moment by an individual creature. The idea is that a creature would pursue proximate

goals, under the guidance of valuation mechanisms, but over long stretches of time, these proximate mechanisms serve improved reproductive success or they don't persist. Adam Smith had similar ideas about the individual pursuit of wealth and referred to the overriding guidance of the "invisible hand." See Smith, 1991.

165 *it produces useful hypotheses about goals that inform the style and structure of formal models of emotions, especially reinforcement learning models* Cohen, 2005.

167 *Brain imaging experiments have now begun to probe the expression and repayment of trust across two interacting humans* McCabe et al., 2001; King-Casas et al., 2005.

167 *Functional MRI uses the MRI device in a different mode of operation and instead of static snapshots, it makes "movies" of microscopic blood-flow changes in your brain* The functional MRI technique was pioneered in the early 1990s: Ogawa et al., 1990ab, 1992, 1993; Kwong et al., 1992. There are now numerous books on the subject, each emphasizing different aspects of the method—Jezzard et al., 2001; Buxton, 2002; Huettel et al., 2003. Raichle, 2003, gives an update of the current status of the field, and the work of Nikos Logothetis and colleagues has provided detailed insight into the relationship between the fMRI "signal" and underlying neural events: Logothetis et al., 2001, 2002; Logothetis, 2003; Logothetis and Pfeuffer, 2004.

A singular effort in providing analysis tools in the neuroimaging domain has been carried out by Karl Friston and Richard Frackowiak at the functional imaging lab at University College, London. This group continues to provide a valuable resource to the imaging community (e.g., Friston et al., 1990, 1991, 1995a, 1995b).

168 *This kind of experiment has now been carried out across the continental United States and in similar experiments between the United States and China and Germany* These kinds of experiments are ongoing between the Human Neuroimaging Lab at Baylor College of Medicine in Houston, Texas, and institutions in Ulm, Germany; Hong Kong; and Guangzhou, China. The basic idea is to probe the neural responses underlying cultural differences in the trust task. One particularly important issue relates to "in group"–"out group" distinctions; a categorization easily made when one knows one is playing someone from a different culture. See King-Casas et al., 2005, for the basic results within culture. Brooks King-Casas, one of the coauthors on that paper, has now taken the two-subject hyperscanned trust game into clinical populations with a perturbed capacity to trust: sufferers of borderline personality disorder. His colleagues Pearl Chiu and Amin Kayali are executing the same tactic and using the trust game on subjects with a perturbed capacity to build and respond to social models of others—autistic and Asperger's patients.

168 *This is a hyperscan experiment, and the participants are half a continent apart in Houston and Pasadena, playing a game, while their brains are scanned* These experiments were carried out in collaboration with two groups at Caltech in Pasadena, California: one headed by cognitive neuroscientist Steve Quartz

(Quartz and Sejnowski, 2002) and the other headed by behavioral economist Colin Camerer (Camerer, 2003). The software that makes the hyperscanning experiments possible was developed at Baylor College of Medicine by the author's group. An early prototype of this system was developed in collaboration with Greg Berns at Emory University. The software, called NEMO, is freely available and can be found at http://www.hnl.bcm.tmc.edu/nemo/. Quartz was instrumental in convincing my group that social cognition was the next great frontier for computational neuroscience and for novel imaging approaches.

169   *yet trust is now being probed in behavioral and imaging experiments all over the globe* Camerer, 2003, reviews trust games across a variety of cultures and settings.

170   *who cannot build proper social models of other humans, misattribute the intentions of others, and are generally unable to trust others* Versions of multiround trust games are now being used to probe subjects afflicted with various personality disorders and with pervasive developmental disorders like autism and Asperger's syndrome. This work should provide "brain response phenotypes" associated with these disorders that can help give insight into these hard-to-understand disorders.

171   *More remarkably, this "intention to increase trust" signal had all the characteristics of a reward-prediction error signal* Compare the predicted signals in Montague et al., 1996, and Schultz et al., 1997, to those observed in King-Casas et al., 2005. In this task, the "intention to trust" is an abstract plan to do something in the near future. This outcome provides evidence for how abstract ideas gain the behavioral potency of primary reinforcers: They act like reward signals at the level of the midbrain and striatum, structures that we share with every vertebrate on the planet. So how do ideas veto instincts? They momentarily gain more importance than instinct through their access to "lower level" reward-harvesting machinery. This is what gives ideas their "meaning."

172   *There were several technical issues involved, including the time difference with China and possibilities for larger-than-normal Internet delays* The exact delay is not terribly crucial, provided that it's under one second. fMRI signals take multiple seconds to build to a peak. The problem is that the experiments can't run properly if the connection delays are too variable.

172   *Paul Chu, the president of Hong Kong University of Science and Technology (HKUST), pioneer in superconductivity, and recipient of the National Medal of Science* Dr. Chu was instrumental in helping get this effort on its feet. Without his "top down" help, I suspect that I could not have gotten the attention of the right individuals.

173   *joint effort between the hyperscan development team along with several students and postdocs in my lab, and a collection of terrific collaborators in China* The Chinese end of the Hong Kong collaboration has been spearheaded by Professor Soo Hong Chew and Professor Rami Zwick. Their boldness in this collaboration has been remarkable, a nonstandard collaboration between professors

in a business school at Hong Kong University of Science and Technology and a science unit at a private medical college in the United States (Baylor College of Medicine).

175 *a self-help manual about how to transform "the road not taken" into real opportunities* Roese, 2005. Also see Roese and Summerville, 2005.

175 *early 1980s by Graham Loomes and Robert Sugden, and separately by Daniel Bell* Loomes and Sugden, 1982; Bell, 1982. These ideas were originally proposed to account for "anomalies" in decision-making. Prospect theory, the result of Amos Tversky and Daniel Kahneman's work, also accounted for a wide range of decision-making anomalies and was a more transparent account of the way that human choice deviates from the predictions of rational choice theory. I am not equipped to give an account of the sociology of this area; however, regret-theoretical accounts allowed for nontransitivity of preferences, which may have dampened enthusiasm for this theoretical perspective.

178 *use some standard models that produce price changes that "look like" real price changes in real markets* These are called drift and diffuse models, basically models with two parts: a trend (the drift part) plus noise (the diffuse part).

178 *we planned to carry out our next experiment using a model of market bubbles conceived by Economics Nobel Laureate Vernon Smith* Caginalp et al., 2001.

181 *In fact, as cultural and financial historians run that period through various analyses, regret about a host of decisions, individual and institutional, permeates every one* Bernanke, 2000; Smiley, 2003.

183 *One such game is called the ultimatum game* Guth et al., 1982.

184 *This kind of experiment has been carried out across many different cultures and for different levels of stakes* Camerer, 2003.

184 *offers at a cost to themselves—the "leave it" option where everyone walks away with nothing* Fehr and Simon, 1999; Fehr and Simon, 2000; Fehr and Rockenbach, 2004; Fehr and Fishbacher, 2004. But see Trivers, 2002, and Trivers, 2004, for a different view on social reciprocity.

185 *show directly how brain responses correlated with the degree of fairness of the offers made by the proposer* Sanfey et al., 2003.

186 *What you did to me today was coming back to you tomorrow in kind* Trivers, 2002.

186 *What about specific reactions to fairness or unfairness? How willing are we to act to impose a cost on unfair behavior?* See Montague et al., 2006, for review of imaging experiments on this topic. The evolution of interaction instincts is a difficult problem. Nowak and Sigmund, 2005, explain them in terms of the future value of reputation. The propensity of humans to enforce norms of behavior even at a cost to themselves (so-called altruistic punishment) is hypothesized to have strengthened social cohesion in times of serious environmental challenges (Fehr and Simon, 2000; Fehr and Fishbacher, 2003).

189 *The father of the modern idea of reciprocal altruism, Robert Trivers, remains skeptical of this interpretation* Trivers, 2004. In science, ideas without

skeptics are like boats with no water on which to float. The bigger and smarter one's skeptic, the more likely it is that your idea or your finding is important—maybe wrong—but important nevertheless. Despite this skepticism, Fehr's results are generated from exceptionally well-controlled experiments.

189 *The idea of inequality aversion is that I dislike a $95/$5 split favoring me, but I also dislike a $5/$95 favoring you, although I dislike the split favoring you a little more* Humans do not like unequal splits across interacting individuals. Fehr and Schmidt, 1999, interpret this in terms of inequality aversion. Their formal expression is given in the text. Inequality aversion is not restricted to humans; monkeys don't like inequality, either, as the work of Sarah Brosnan and Franz de Waal has shown (Brosnan and de Waal, 2003; Brosnan et al., 2005). Tania Singer and colleagues (2006) have recently shown that these fairness instincts about exchange compile into our perception of others' pain, that is, if I see you being unfair to someone, then I will tolerate watching you experience more pain than I otherwise would. Not surprisingly, the effect is seen most strongly in males, and the researchers have suggested that this may align with this historical role of males as enforcers of social norms.

192 *may lead further to a real alliance between economists and neuroscientists* Fehr's experimental results have been so clear that they have forced behavioral economists, sociobiologists (Trivers), and neuroscientists to respond to them. I think that they provide a touchstone area ripe for making theoretical strides that connect neural mechanism with evolutionary explanation. The neuroscientists, including myself, have tended to focus on learning theories, while the economists have tended to focus on equilibrium explanations—that is, ultimate evolutionary reasons for the character of social mechanisms.

193 *an invigorating account of why cultures advance at vastly different rates* Diamond, 1997.

194 *facts highlighted by Diamond and analyzed by others* E.g., Maisels, 1990.

197 *Vampire bats share blood, chimps share meat, and humans share nearly everything, but especially information* Wilkinson, 1984, 1990.

197 *economic exchange may have been the feature that separated the early modern humans from Neanderthals* Horan et al., 2006, offer the following in the abstract of their recent paper:

> One of the great puzzles in science concerns the rise of early modern humans and the fall of Neanderthals. A number of theories exist and many support the biological principle of competitive exclusion: If two similar species occupy exactly the same niche, only the most efficient will survive; the other will go extinct. Such ideas of biological efficiency pertain to biological or physiological factors like lower mortality rates or greater efficiency in hunting. Evidence for such mechanistic theories in which biology is destiny, however, is

limited. In response, this paper develops a behavioral model of Neanderthal extinction. We show how the division of labor and subsequent trading among early modern humans could have helped
them to overcome potential biological deficiencies, and therefore
lead to the demise of Neanderthals.

The behavioral theory they propose would indeed have to rely upon brain
mechanisms inducing fair trade. And as I have discussed throughout this
book, such mechanisms would be expected in the presence of constraints that
enforce efficient decisions and actions.

CHAPTER 7: FROM PEPSI TO TERRORISM

199   *Two researchers wrote this wonderful passage in the journal* Neuron Anderson
      and Sobel, 2003.

202   *visual system scientists, have exploited illusions for many years to probe the way
      important perceptions are constructed* Gregory, 2006.

202   *an effect is perceived to occur before a cause, demonstrating a remarkable flexibil-
      ity in our mental software that has only recently been identified* This remarkable
      illusion, discovered recently by David Eagleman, is particularly important,
      since it shows that even our "intuitive" sense of causality, and logic based on it,
      are themselves constructs.

204   *advertisements and games pitched at children. Are these capable of damaging
      normal reward-processing and choice mechanisms?* Koepp et al., 1998.

204   *you sick or causes a bad experience, and, poof!—good-bye brand value—
      especially if it makes you sick* Recent events in the news make this assertion
      quite real. Remember the claim that a severed finger was found in a bowl of
      chili at a well-known fast-food restaurant. The claim ended up being proved
      a fake, but the psychological and hence economic consequences were real
      and measurable.

205   *has assigned social status to sports cars, and this status generates measurable re-
      sponses in brains* Erk et al., 2002.

208   *a region just above and between your eyes (but of course inside your skull)* Mc-
      Clure et al., 2004.

210   *to art by asking subjects placed in functional MRI machines to rank art as "ugly,"
      "neutral," or "beautiful"* Kawabata and Zeki, 2004. Also see Aharon et al., 2001.

211   *strongly support the idea that the brain must place diverse stimuli on common
      valuation scales* Glimcher, 2003.

211   *an extremely diverse range of information sources to compute common internal
      currencies—common valuation scales* See Bechara et al., 2005, for a review of
      the somatic marker hypothesis. See Montague and Berns, 2002, for a discus-
      sion of common currency ideas.

215   *"like" pollen and nectar, and the learning machinery in its brain is equipped to*

assign value to other stimuli that predict either of the prewired rewards Menzel
et al., 2006; Hammer, 1997.

217   *humans has already inspired funding agencies to wonder about using the*
models to understand drug and gambling addictions The National Institute
on Drug Abuse (NIDA) has sponsored conferences and grants to use computational models to understanding the algorithmic underpinnings of drug
and gambling addictions. I think that they have shown great foresight in
these efforts.

CHAPTER 8: OUR CHOICE

221   *computational complexity with more much finesse in order to distinguish the*
person from the tomato—current methods are far too blunt to deal with this
problem Sipser, 2005.

227   *Author Tom Wolfe portrayed his reaction to the sister sciences of mind and brain*
Wolfe, 1996.

228   *semiquantitative, making predictions that agree quantitatively with some data*
and qualitatively with the rest Montague et al., 2006, review the way these
models have been used in human neuroimaging experiments. Their semiquantitative nature is highlighted here.

228   *forward not backward, real progress is being made on many fronts, and things*
are indeed accelerating The advancement is clear in the progression of books
on the subject—Churchland and Sejnowski, 1992; Rieke et al., 1997; Koch,
1999; Dayan and Abbott, 2001.

230   *Daniel Dennett addresses these questions in* Freedom Evolves Dennett, 2003.

232   *It's reasonable to suspect that this transition correlates with a transition of neu-*
ral activation from the prefrontal cortex to the striatum Magill, 2003.

241   *capacity of the prefrontal cortex to participate in the sustenance of new instinct-*
vetoing ideas that act to inhibit impulses Anderson et al., 1999.

245   *experiments in animals have not separated clearly the influence of attention*
and the influence of reward Maunsell, 2004. This paper points out that
work on attention carried out in animals has not separated clearly the neural responses to reward and to attention. One problem is that the word *at-*
tention is used liberally in the neuroscience literature, as is the word
reward. Although I have discussed formal models of reward processing,
not all neuroscientists would restrict their use of the word *reward* in the
way that I have. The same general issue arises with the word *attention*,
which is generally used as synonymous with increased sensitivity along
some sensory dimension. As suggested in the text, I suspect that in the future workable theories of attention may indeed look like resource distribution strategies and will overlap substantially with reinforcement learning
theories of reward processing. There may even emerge new concepts as this
work progresses.

246   *but if the animal were carrying out a visual task, then her model would also capture properties of visual attention* Niv, 2005.

EPILOGUE: ARE HUMANS COMPUTABLE?

254   *uncomputable to account for the totality of our mental operations; he hypothesizes a physical basis for the uncomputable parts* Penrose, 1989, 1994.

Bibliography

Abbott, L. F. 2003. Balancing homeostasis and learning in neural circuits. *Zoology* 106(4):365–371.

Abbott, L. F., and W. G. Regehr. 2004. Synaptic computation. *Nature* 431:796–803.

Ackley, D. H., G. E. Hinton, and T. J. Sejnowski. 1985. A learning algorithm for Boltzmann machines. *Cognitive Science* 9:147–169.

Adams, D. *The Hitchhiker's Guide to the Galaxy*. 1980. New York: Ballantine Books.

Aharon, I., N. Etcoff, D. Ariely, C. F. Chabris, E. O'Connor, and H. C. Breiter. 2001. Beautiful faces have variable reward value: fMRI and behavioral evidence. *Neuron* 32(3):537–551.

Allen, C., and C. F. Stevens. 1994. An evaluation of causes for unreliability of synaptic transmission. *Proceedings of the National Academy of Sciences (USA)* 91: 10380–10383.

Anderson, A. K., and N. Sobel. 2003. Dissociating intensity from valence as sensory inputs to emotion. *Neuron* 39:581–583.

Anderson, C. H., D. C. Van Essen, and B. A. Olshausen. 2005. Directed visual attention and the dynamic control of information flow. In *Neurobiology of Attention*, eds. L. Itti, G. Rees, J. Tsotsos. Academic Press/Elsevier, pp. 11–17.

Anderson, S. W., A. Bechara, H. Damasio, D. Tranel, and A. R. Damasio. 1999. Impairment of social and moral behavior related to early damage in human prefrontal cortex. *Nature Neuroscience* 2(11):1032–1037.

Aston-Jones, G., and J. D. Cohen. 2005. An integrative theory of locus coeruleus-norepinephrine function: adaptive gain and optimal performance. *Annual Review of Neuroscience* 28:403–450.

Aston-Jones, G., J. Rajkowski, P. Kubiak, and T. Alexinsky. 1994. Locus coeruleus neurons in monkey are selectively activated by attended cues in a vigilance task. *Journal of Neuroscience* 14(7):4467–4480.

Attwell, D., and S. B. Laughlin. 2001. An energy budget for signaling in the grey matter of the brain. *J Cerebral Blood Flow Metabolism* 21(10):1133–1145.

Axelrod, R. M. 1984. *The Evolution of Cooperation*. New York: Basic Books.

Axelrod, R. 1986. An evolutionary approach to norms. *American Political Science Review* 80:1095–2011.

Axelrod, R., and W. D. Hamilton. 1981. The evolution of cooperation. *Science* 211:1390–1396.

Ballard, D. H., G. E. Hinton, and T. J. Sejnowski. 1983. Parallel visual computation. *Nature* 306:21–26.

Barkow, J. H., L. Cosmides, and J. Tooby. 1992. *The Adapted Mind*. New York: Oxford University Press.

Barto, A. G., R. S. Sutton, and C. W. Anderson. 1983. Neuronlike adaptive elements that can solve difficult learning control problems. *IEEE Transactions on Systems, Man, and Cybernetics* 13:834–846.

Bayer, H. M., and P. W. Glimcher. 2005. Midbrain dopamine neurons encode a quantitative reward prediction error signal. *Neuron* 47:129–141.

Bear, M. F., B. W. Conners, and M. A. Paradiso. 1996. *Neuroscience: Exploring the Brain*. New York: Williams and Wilkins.

Bechara, A., H. Damasio, D. Tranel, and A. R. Damasio. 2005. The Iowa Gambling Task and the somatic marker hypothesis: some questions and answers. *Trends in Cognitive Science* 9(4):159–162.

Bell, D. 1982. Regret in decision making under uncertainty. *Operations Research* 30:961–981.

Belliveau, J., et al. 1991. Functional mapping of the human visual cortex by magnetic resonance imaging. *Science* 254:716–719.

Belliveau, J., B. Rosen, et al. 1990. Functional cerebral imaging by susceptibility-contrast NMR. *Magnetic Resonance in Medicine*. 14:538–546.

Benioff, P. 1980. The computer as a physical system: A microscopic quantum mechanical Hamiltonian model of computers as represented by Turing machines. *Journal of Statistical Physics* 22:563–591.

Benioff, P. 1982. Quantum mechanical Hamiltonian models of Turing machines. *Journal of Statistical Physics* 29:515–546.

Bennett, C. H. 1973. Logical reversibility of computation. *IBM Journal of Research and Development*. 17:525–532.

Bennett, C. H. 1979. Dissipation-error tradeoff in proofreading. *BioSystems* 11:85–91.

Bennett, C. H. 1982. The thermodynamics of computation. *International Journal of Theoretical Physics*. 21:905–940.

Bennett, C. H., et al. 1993. Teleporting an unknown quantum state via dual classical and Einstein-Podolsky-Rosen channels. *Physical Review Letters* 70:1895–1899.

Bennett, C. H., et al. 1994. Reduction of quantum entropy by reversible extraction of classical information. *Journal of Modern Optics* 41:2307–2314.

Bennett, C. H., P. Gács, M. Li, P.M.B. Vitányi, and W. Zurek. 1998. Thermodynamics of computation and information distance. *Proc. 1993 ACM Symposium on Theory of Computing,* and *IEEE Transactions on Information Theory.* IT-44:4(1998) 1407–1423.

Bennett, C. H., and R. Landauer. 1985. Fundamental physical limits of computation. *Scientific American* 253:48–56. (July issue.)

Bennett, C. H., and S. J. Wiesner. 1992. Communication via 1- and 2-particle operators on Einstein-Podolsky-Rosen states. *Physical Review Letters* 69:2881–2884.

Berg, H. 2004. *E. Coli in Motion.* New York: Springer-Verlag.

Berg, J., J. Dickhaut, and K. McCabe. 1995. Trust, reciprocity, and social history. *Games and Economic Behavior* 10:122–142.

Bernanke, B. S. 2000. *Essays on the Great Depression.* Princeton: Princeton University Press.

Bernheim, B. D., and A. Rangel. 2004. Addiction and cue-triggered decision processes. *American Economic Review* 94(5):1558–1590.

Berns, G. S. 2005. *Satisfaction.* New York: Holt.

Bertsekas, D. P. 2003. *Convex Analysis and Optimization.* Belmont, Mass.: Athena Scientific.

Bertsekas, D. P., and J. N. Tsitsiklis. 1996. *Neuro-Dynamic Programming.* Belmont, Mass.: Athena Scientific.

Bialek, W. 1987. Physical limits to sensation and perception. *Annual Review of Biophysics and Biophysical Chemistry* 16:455–478.

Bialek, W., et al. 1991. Reading a neural code. *Science* 252:1854–1857.

Bialek, W., and W. G. Owen. 1990. Temporal filtering in retinal bipolar cells. Elements of an optimal computation? *Biophysical Journal* 58(5):1227–1233.

Bialek, W., and F. Reike. 1992. Reliability and information transmission in spiking neurons. *Trends in Neuroscience* 15(11):428–434.

Bialek, W., and A. Schweitzer. 1985. Quantum noise and the threshold of hearing. *Physical Review Letters* 54(7):725.

Bialek, W., and S. Setayeshgar. 2005. Physical limits to biochemical signaling. Proceedings of the *National Academy of Sciences (USA)* 102(29):10040–10045.

Bialek, W., and A. Zee. 1988. Understanding the efficiency of human perception. *Physical Review Letters* 61(13):1512–1515.

Braitenberg, V., and A. Schuz. 1998. *Cortex: Statistics and Geometry of Neuronal Connectivity.* New York: Springer.

Breiter, H. C., I. Aharon, D. Kahneman, A. Dale, and P. Shizgal. 2001. Functional imaging of neural responses to expectancy and experience of monetary gains and losses. *Neuron* 30:619–639.

Britton, J. L., D. C. Ince, and P. T. Saunders. 1992. *The Collected Works of A. M. Turing.* New York: Elsevier.

Brody, C. D., A Hernández, A. Zainos, and R. Romo. 2003. Timing and neural encoding of somatosensory parametric working memory in macaque prefrontal cortex. *Cerebral Cortex* 13(11):1196–1207.

Brosnan, S. F., and F. B. de Waal. 2003. Monkeys reject unequal pay. *Nature* 425: 297–299.

Brosnan, S. F., H. C. Schiff, and F. B. de Waal. 2005. Tolerance for inequity may increase with social closeness in chimpanzees. *Proceedings of the Royal Society of London B* 272:253–258.

Bush, R. R., and F. Mosteller. 1955. *Stochastic Models for Learning*. New York: Wiley.

Buss, D. M. 2004. *Evolutionary Psychology: The New Science of the Mind*. 2nd ed. Boston: Allyn & Bacon.

Buxton, R. B. 2002. *Introduction to Functional Magnetic Resonance Imaging*. Cambridge, UK: Cambridge University Press.

Caginalp, G., D. Porter, and V. Smith. 2001. Financial bubbles: Excess cash, momentum, and incomplete information. *Journal of Psychology and Financial Markets* 2(2):80–99.

Calvin, W. 1993. *Tools, Language, and Cognition in Human Evolution*, ed. Kathleen R. Gibson and Tim Ingold. Cambridge: Cambridge University Press, pp. 230–250.

Camerer, C. F. 2003. *Behavioral Game Theory*. Princeton: Princeton University Press.

Camerer, C. F., and E. Fehr. 2006. When does "economic man" dominate social behavior? *Science* 311:47–52.

Camerer, C. F., and R. H. Thaler. 1995. Ultimatums, dictators, and manners. *Journal of Economic Perspectives*. 9:209–219.

Chaitin, G. J. 1987. *Algorithmic Information Theory*. Cambridge: Cambridge University Press.

Chalmers, D. 1996. *The Conscious Mind*. Oxford: Oxford University Press.

Cherniak, C. 1992. Local optimization of neuron arbors. *Biological Cybernetics* 66(6):503–510.

Cherniak, C. 1994. Component placement optimization in the brain. *Journal of Neuroscience* 14:2418–2427.

Cherniak, C. 1995. Neural component placement. *Trends in Neuroscience* 18(12): 522–527.

Cherniak, C., Z. Mokhtarzada, R. Rodriguez-Estéban, and K. Changizi. 2004. Global optimization of cerebral cortex layout. *Proceedings of the National Academy of Sciences (USA)* 101(4):1081–1086.

Chklovskii, D. B., and A. A. Koulakov, 2000. A wire length minimization approach to ocular dominance patterns in mammalian visual cortex. *Physica A* 284: 318–334.

Chklovskii, D. B., and A. A. Koulakov. 2004. Maps in the brain: What can we learn from them? *Annual Review of Neuroscience* 27:369–392.

Chklovskii, D. B., B. W. Mel, and K. Svoboda. 2004. Cortical rewiring and information storage. *Nature* 431:782–788.

Chklovskii, D. B., T. Schikorski, and C. F. Stevens. 2002. Wiring optimization in cortical circuits. *Neuron* 34(3):341–347.

Churchland, P. M. 1981. Eliminative materialism and the propositional attitudes.

Journal of Philosophy 78(2):67–90. Reprinted in *Mind and Cognition*, ed. W. G. Lycan. Oxford: Blackwell, 1999.

Churchland, P. M. 1984. *Matter and Consciousness*. Cambridge, Mass.: MIT Press.

Churchland, P. M. 1995. *The Engine of Reason: The Seat of the Soul*. Cambridge, Mass.: MIT Press.

Churchland, P. S. 1982. Mind-brain reduction: new light from the philosophy of science. *Neuroscience* 7(5):1041–1047.

Churchland, P. S. 1986. *Neurophilosophy*. Cambridge, Mass.: MIT Press.

Churchland, P. S. 1988. The significance of neuroscience for philosophy. *Trends in Neuroscience* 11(7):304–307.

Churchland, P. S. 1998. What should we expect from a theory of consciousness? *Advances in Neurology* 77:19–30.

Churchland, P. S. 2002. Self-representation in nervous systems. *Science* 296:308–310.

Churchland, P. S., and P. M. Churchland. 2002. Neural worlds and real worlds. *Nature Reviews Neuroscience* 3(11):903–907.

Churchland, P. S., and T. J. Sejnowski. 1988. Perspectives on cognitive neuroscience. *Science* 242:741–745.

Churchland, P. S., and T. J. Sejnowski. 1992. *The Computational Brain*. Cambridge, Mass.: MIT Press.

Cohen, J. D. 2005. The vulcanization of the human brain: A neural perspective on interactions between cognition and emotion. *Journal of Economic Perspectives* 19:3–24.

Cohen, J. D., T. S. Braver, and J. W. Brown. 2002. Computational perspectives on dopamine function in prefrontal cortex. *Current Opinion in Neurobiology* 12:223–229.

Cohen, J. D., and D. Servan-Schreiber. 1992. Context, cortex, and dopamine: A connectionist approach to behavior and biology in schizophrenia. *Psychological Review* 1:45–77.

Collins, S. H., A. Ruina, R. Tedrake, and M. Wisse. 2005. Efficient bipedal robots based on passive-dynamic walkers. *Science* 307:1082–1085.

Coricelli, G., H. D. Critchley, M. Joffily, J. P. O'Doherty, A. Sirigu, and R. J. Dolan. 2005. Regret and its avoidance: A neuroimaging study of choice behavior. *Nature Neuroscience* 8(9):1255–1262.

Damasio, A. R. 1994. *Descartes' Error*. New York: G. P. Putnam's Sons.

Damasio, A. R. 1999. *The Feeling of What Happens*. Orlando: Harcourt.

Dayan, P. 1992. The convergence of TD(λ) for general lambda. *Machine Learning* 8:341–362.

Dayan, P. 1993. Improving generalisation for temporal difference learning: The successor representation. *Neural Computation* 5:613–624.

Dayan, P., and L. F. Abbott. 2001. *Theoretical Neuroscience*. Cambridge, Mass.: MIT Press.

Dayan, P., S. Kakade, and P. R. Montague. 2000. Learning and selective attention. *Nature Neuroscience* 3:1218–1223.

Dayan, P., and T. J. Sejnowski. 1994. TD(lambda) converges with probability 1. *Machine Learning* 14:295–301.

Dayan, P., and C.J.C.H. Watkins. 2001. Reinforcement learning. *Encyclopedia of Cognitive Science.* London: MacMillan Reference, Ltd.

Daw, N. D. 2003. Reinforcement learning models of the dopamine system and their behavioral implications. Ph.D. thesis. Carnegie Mellon University, Department of Computer Science.

Daw, N. D., Y. Niv, and P. Dayan. 2005. Uncertainty-based competition between prefrontal and dorsolateral striatal systems for behavioral control. *Native Neuroscience* 8: 1704–1711.

Dawkins, R. 1976. *The Selfish Gene.* Oxford: Oxford University Press.

Dawkins, R. 1982. *The Extended Phenotype.* Oxford: Oxford University Press.

Dawkins, R. 1995. God's utility function. *Scientific American* (November issue), pp. 80–85.

Delgado, M. R., R. H. Frank, and E. A. Phelps. 2005. Perceptions of moral character modulate the neural systems of reward during the trust game. *Nature Neuroscience* 8:1611–1618.

Delgado, M. R., L. E. Nystrom, C. Fissell, D. C. Noll, and J. A. Fiez. 2000. Tracking the hemodynamic responses to reward and punishment in the striatum. *Journal of Neurophysiology* 84:3072–3077.

Deneve, S., and A. Pouget. 2003. Basis functions for object-centered representations. *Neuron* 37(2):347–359.

Dennett, D. C. 1991. *Consciousness Explained.* Boston: Little, Brown & Company.

Dennett, D. C. 1995. *Darwin's Dangerous Idea.* New York: Touchstone.

Dennett, D. C. 2003. *Freedom Evolves.* New York: Penguin.

de Quervain, D. J., U. Fischbacher, V. Treyer, M. Schellhammer, and U. Schnyder, et al. 2004. The neural basis of altruistic punishment. *Science* 305:1254–1258.

Deutsch, D. 1985. Quantum theory, the Church-Turing principle and the universal quantum computer. *Proceedings of the Royal Society of London A* 400:97–119.

Deutsch, D. 1997. *The Fabric of Reality.* New York: Penguin Books.

Deutsch, D. 1999. Quantum theory of probability and decisions. *Proceedings of the Royal Society of London A* 455:3129–3137.

Devarajan, S., and A. C. Fisher. 1981. Hotelling's "Economics of Exhaustible Resources": Fifty years later. *Journal of Economic Literature* 19:65–73.

Diamond, J. 1997. *Guns, Germs, and Steel.* New York: W. W. Norton and Co.

Dirac, P. 1930. *Principles of Quantum Mechanics.* London: Oxford University Press.

Dobzhansky, T. 1973. Nothing in biology makes sense except in the light of evolution. *The American Biology Teacher* 35:125–129.

Dolan, R. J., P. Fletcher, C. D. Frith, K. J. Friston, R. S. Frackowiak, and P. M. Grasby. 1995. Dopaminergic modulation of impaired cognitive activation in the anterior cingulate cortex in schizophrenia. *Nature* 378: 180–182.

Dolan, R. J., and K. J. Friston. 1989. Positron emission tomography in psychiatric and neuropsychiatric disorders. *Seminars in Neurology* 9(4):330–337.

Dönitz, Karl. 1997. *Memoirs: Ten Years and Twenty Days.* Annapolis: Da Capo Press.

Dreyfus, H. 1972. *What Computers Can't Do.* New York: Harper and Row.

Dreyfus, H. 1992. *What Computers Still Can't Do.* Cambridge, Mass.: MIT Press.

Durbin, R., and G. Mitchison. 1990. A dimension reduction framework for understanding cortical maps. *Nature* 343:644–647.

Durstewitz, D., J. K. Seamans, and T. J. Sejnowski. 2000. Neurocomputational models of working memory. *Nature Neuroscience* 3:1184–1191.

Dusenbery, D. B. 1992. *Sensory Ecology*. New York: W. H. Freeman and Co.

Dusenbery, D. B. 1997. Minimum size limit for useful locomotion by free-swimming microbes. *Proceedings of the National Academy of Sciences (USA)* 94:10949–10954.

Edelman, G. M., and G. Tononi. 2000. *A Universe of Consciousness*. New York: Basic Books.

Egelman, D. M., C. Person, and P. R. Montague. 1998. A computational role for dopamine delivery in human decision-making. *Journal of Cognitive Neuroscience* 10:623–630.

Eliasmith, C. 2005. A unified approach to building and controlling spiking attractor networks. *Neural Computation* 17(6):1276–1314.

Eliasmith, C., and C. H. Anderson. 2002. *Neural Engineering*. Cambridge, Mass.: MIT Press.

Erk, S., M. Spitzer, A. P. Wunderlich, L. Galley, and H. Walter. 2002. Cultural objects modulate reward circuitry. *Neuroreport* 13:2499–2503.

Fahlman, S. E., G. E. Hinton, and T. J. Sejnowski. 1983. Massively parallel architectures for A. I.: Netl, Thistle, and Boltzmann machines. *Proceedings of the National Conference on Artificial Intelligence*, Washington, D.C.

Fairhall, A. L., G. D. Lewen, W. Bialek, and R. R. de Ruyter van Steveninck. 2001. Efficiency and ambiguity in an adaptive neural code. *Nature* 412:787–792.

Faisal, A. A., J. A. White, and S. B. Laughlin. 2005. Ion-channel noise places limits on the miniaturization of the brain's wiring. *Current Biology* 15(12):1143–1149.

Farhi, E., J. Goldstone, S. Gutmann, A. Lapan, A. Lundgren, and D. Preda. 2001. Quantum adiabatic evolution algorithm applied to random instances of an NP-complete problem. *Science* 292:472–476.

Fehr, E., and U. Fischbacher. 2003. The nature of human altruism. *Nature* 425:785–791.

Fehr, E., and U. Fischbacher. 2004. Social norms and human cooperation. *Trends in Cognitive Science* 8:185–189.

Fehr, E., and S. Gächter. 2002. Altruistic punishment in humans. *Nature* 415:137–140.

Fehr, E., and B. Rockenbach. 2004. Human altruism: Economic, neural, and evolutionary perspectives. *Current Opinion in Neurobiology* 14:784–790.

Fehr, E., and K. M. Schmidt. 1999. A theory of fairness, competition, and cooperation. *Quarterly Journal of Economics*. 71:397–404.

Fehr, E., and G. Simon. 2000. Fairness and retaliation: The economics of reciprocity. *Journal of Economic Perspectives* 14(3):159–181.

Felleman, D., and D. Van Essen. 1991. Distributed hierarchical processing in the primate cerebral cortex. *Cerebral Cortex* 1:1–47.

Fodor, J. 1975. *The Language of Thought*. New York: Thomas Crowell.

Fox, P. T., M. A. Mintun, M. E. Raichle, and P. Herscovitch. 1984. A noninvasive approach to quantitative functional brain mapping with H2 (15)O and positron emission tomography. *Journal of Cerebral Blood Flow and Metabolism* 4(3):329–333.

Fox, P. T., and M. E. Raichle. 1986. Focal physiological uncoupling of cerebral blood flow and oxidative metabolism during somatosensory stimulation in human subjects. *Proceedings of the National Academy of Sciences (USA)* 83(4):1140–1144.

Fox, P. T., M. E. Raichle, M. A. Mintun, and C. Dence. 1988. Nonoxidative glucose consumption during focal physiologic neural activity. *Science* 241:462–464.

Fredkin, E., and T. Toffoli. 1982. Conservative logic. *International Journal of Theoretical Physics* 21:219–253.

Friston, K. J., C. D. Frith, P. F. Liddle, R. J. Dolan, A. A. Lammertsma, and R. S. Frackowiak. 1990. The relationship between global and local changes in PET scans. *Journal of Cerebral Blood Flow and Metabolism* 10(4):458–466.

Friston, K. J., C. D. Frith, P. F. Liddle, and R. S. Frackowiak. 1991. Comparing functional (PET) images: The assessment of significant change. *Journal of Cerebral Blood Flow and Metabolism* 1991 Jul; 11(4):690–699.

Friston, K. J., C. D. Frith, R. Turner, R.S.J. Frackowiak. 1995a. Characterising dynamic brain responses with fMRI: A multivariate approach. *NeuroImage* 2: 166–172.

Friston, K. J., A. P. Holmes, K. Worsley, J. B. Poline, C. D. Frith, and R.S.J. Frackowiak. 1995b. Statistical parametric maps in functional brain imaging: A general linear approach. *Human Brain Mapping* 2:189–210.

Friston, K. J., G. Tononi, G. N. Reeke, O. Sporns, and G. M. Edelman. 1994. Value-dependent selection in the brain: Simulation in a synthetic neural model. *Neuroscience* 59:229–243.

Frith, C. D., K. J. Friston, P. F. Liddle, R.S.J. Frackowiak. 1991. A PET study of word finding. *Neuropsychologia* 29(12):1137–1148.

Fuster, J. M. 1980. *The Prefrontal Cortex*. New York: Raven Press.

Gilovich, T., R. Vallone, and A. Tversky. 1985. The hot hand in basketball: On the misperception of random sequences. *Cognitive Psychology* 17:295–314.

Gisiger, T., M. Kerszberg, and J. P. Changeux. 2005. Acquisition and performance of delayed-response tasks: a neural network model. *Cerebral Cortex* 15(5):489–506.

Glimcher, P. W. 2002. Decisions, decisions, decisions: Choosing a biological science of choice. *Neuron* 36(2):323–332.

Glimcher, P. W. 2003. *Decisions, Uncertainty, and the Brain*. Cambridge, Mass.: MIT Press.

Glimcher, P. W. 2005. Indeterminacy in brain and behavior. *Annual Review of Psychology* 56:25–56.

Glimcher, P. W., and A. Rustichini. 2004. Neuroeconomics: The consilience of brain and decision. *Science* 306:447–452.

Gold, J. I., and M. N. Shadlen. 2002. Banburismus and the brain: Decoding the relationship between sensory stimuli, decisions, and reward. *Neuron* 36:299–308.

Goldman, M. S., P. Maldonado, and L. F. Abbott. 2002. Redundancy reduction and sustained firing with stochastic depressing synapses. *Journal of Neuroscience* 22(2):584–591.

Goodhill, G. J., K. R. Bates, and P. R. Montague. 1997. Influences on the global structure of cortical maps. *Proceedings of the Royal Society of London B* 264:649–655.

Goodhill, G. J., and T. J. Sejnowski. 1997. A unifying objective function for topographic mappings. *Neural Computation* 9:1291–1303.

Grafton, S. T., J. C. Mazziotta, S. Presty, K. J. Friston, R. S. Frackowiak, and M. E. Phelps. 1992. Functional anatomy of human procedural learning determined with regional cerebral blood flow and PET. *Journal of Neuroscience* 12(7): 2542–2548.

Graham, G. 2005. *Behaviorism.* http://plato.stanford.edu/entries/behaviorism/

Graybiel, A. M. 2005. The basal ganglia: learning new tricks and loving it. *Current Opinion in Neurobiology* 15(6):638–644.

Gregory, R. 2006. *Illusion.* Oxford, UK: Oxford University Press.

Gruber, J., and K. Botond. 2001. Is addiction "rational"? Theory and evidence. *Quarterly Journal of Economics* 116:1261–1305.

Güth, W., R. Schmittberger, B. Schwarze. 1982. An experimental analysis of ultimatum bargaining. *Journal of Economic Behavior and Organization.* 3:367–388.

Hamilton, W. D. 1964a. The genetical evolution of social behaviour. I. *Journal of Theoretical Biology* 7(1): 1–16.

Hamilton, W. D. 1964b. The genetical evolution of social behaviour. II. *Journal of Theoretical Biology* 7(1): 17–52.

Hammer, M. 1997. The neural basis of associative reward learning in honeybees. *Trends in Neuroscience* 20(6):245–252.

Hauser, M. D. 2000. *Wild Minds: What Animals Really Think.* New York: Henry Holt.

Heeger, D. J., and D. Ress. 2002. What does fMRI tell us about neuronal activity? *Nature Neuroscience* 3(2):142–151.

Heeger, D. J., E. P. Simoncelli, and J. A. Movshon. 1996. Computational models of cortical visual processing. *Proceedings of the National Academy of Sciences (USA)* 93(2):623–762.

Henrich, J., and R. Boyd. 2001. Why people punish defectors: Weak conformist transmission can stabilize costly enforcement of norms in cooperative dilemmas. *Journal of Theoretical Biology* 208:79–89.

Henrich, J., R. Boyd, S. Bowles, C. Camerer, E. Fehr, et al. 2001. Cooperation, reciprocity and punishment in fifteen small-scale societies. *American Economic Review* 91:73–78.

Henrich, J., R. Boyd, S. Bowles, C. Camerer, E. Fehr, et al. 2004. *Foundations of Human Sociality: Economic Experiments and Ethnographic Evidence from Fifteen Small-Scale Societies.* Oxford: Oxford University Press.

Herken, R. (ed.) 1988. *The Universal Turing Machine.* Oxford: Oxford University Press.

Herrnstein, R. J., and D. Prelec. 1991. Melioration: a theory of distributed choice. *Journal of Economic Perspectives.* 5:137–156.

Hilgetag, C. C., G. A. Burns, M. A. O'Neill, J. W. Scannell, and M. P. Young. 2000. Anatomical connectivity defines the organization of clusters of cortical areas in the macaque monkey and the cat. *Philosophical Transactions of the Royal Society of London B* 355:91–110.

Hinton, G., and J. Anderson. 1981. *Parallel Models of Associative Memory*. Hillsdale, NJ.: Lawrence Erlbaum Assoc.

Hinton, G. E., J. L. McClelland, and D. E. Rumelhart. 1986. Distributed representations. In Rumelhart, D. E., and J. L. McClelland, editors, *Parallel Distributed Processing: Explorations in the Microstructure of Cognition. Volume 1: Foundations*. Cambridge, Mass.: MIT Press.

Hodges, A. 2000. *Alan Turing: The Enigma*. New York: Walker and Company.

Hodges, A. 2002. http://www.turing.org.uk/turing/Turing.html—Alan Turing Web page maintained by Andrew Hodges.

Holldobler, B., and E. O. Wilson. 1990. *The Ants*. Cambridge, Mass.: Belknap Press.

Hollerman, J. R., and W. Schultz. 1998. Dopamine neurons report an error in the temporal prediction of reward during learning. *Nature Neuroscience* 1:304–309.

Hopfield, J. J. 1982. Neural networks and physical systems with emergent collective computational abilities. *Proceedings of the National Academy of Sciences (USA)* 79(8):2554–2558.

Hopfield, J. J. 1984. Neurons with graded response have collective computational properties like those of two-state neurons. *Proceedings of the National Academy of Sciences (USA)* 81:3088–3092.

Hopfield, J. J., D. I. Feinstein, and R. G. Palmer. 1983. "Unlearning" has a stabilizing effect in collective memories. *Nature*. 304:158–159.

Hopfield, J. J., and D. W. Tank. 1985. "Neural" computation of decisions in optimization problems. *Biological Cybernetics* 52(3):141–152.

Hopfield, J. J., and D. W. Tank. 1986. Computing with neural circuits: A model. *Science* 233:625–633.

Horan, R. D., E. H. Bulte, and J. F. Shogren. 2006. How trade saved humanity from biological exclusion: The Neanderthal enigma revisited and revised. *Journal of Economic Behavior and Organization* (in press).

Hsu, F. H. 2002. *Behind Deep Blue: Building the Computer that Defeated the World Chess Champion*. Princeton: Princeton University Press.

Huettel, S. A., A. W. Song, and G. McCarthy. 2003. *Functional Magnetic Resonance Imaging*. Sunderland, Mass.: Sinauer Associates.

Huk, A. C., and M. N. Shadlen. 2003. The neurobiology of visual-saccadic decision making. *Annual Review of Neuroscience* 26:133–179.

Hyder, F., D. L. Rothman, and R. G. Shulman. 2002. Total neuroenergetics support localized brain activity: Implications for the interpretation of fMRI. *Proceedings of the National Academy of Sciences (USA)* 99(16):10771–10776.

Hyman, S. E., and R. C. Malenka. 2001. Addiction and the brain: The neurobiology of compulsion and its persistence. *Nature Reviews Neuroscience* 2:695–703.

Hyman, S. E., E. J. Nestler, and R. C. Malenka. 2006. Reward-related memory and addiction. *Annual Reviews in Neuroscience* 29 (in press).

Insel, T. R., and L. J. Young. 2001. The neurobiology of attachment. *Nature Reviews Neuroscience* 2:129–136.

Jezzard, P., P. M. Matthews, and S. M. Smith. 2001. *Functional Magnetic Resonance Imaging*. Oxford: Oxford University Press.

Johnston, D., D. A. Hoffman, C. M. Colbert, and J. C. Magee. 1999. Regulation of back-propagating action potentials in hippocampal neurons. *Current Opinion in Neurobiology* 9(3):288–292.

Johnston, D., and S. M. Wu. 1994. *Foundations of Cellular Neurophysiology*. Cambridge, Mass.: MIT Press.

Jones, A. K., K. Friston, and R. S. Frackowiak. 1992. Localization of responses to pain in human cerebral cortex. *Science* 255(5041):215–216.

Kaelbling, L. P., M. L. Littman, and A. W. Moore. 1996. Reinforcement learning: A survey. *Journal of Artificial Intelligence Research* 4:237–285.

Kagel, J. H., and A. E. Roth. 1995. *The Handbook of Experimental Economics*. Princeton: Princeton University Press.

Kahneman, D. A. 2003. Perspective on judgment and choice: mapping bounded rationality. *American Psychologist* 58(9):697–720.

Kahneman, D. A., and A. Tversky. 1979. Prospect theory: An analysis of decision under risk. *Econometrica* 47:263–291.

Kalivas, P. W., and N. D. Volkow. 2005. The neural basis of addiction: a pathology of motivation and choice. *American Journal of Psychiatry* 162(8):1403–1413.

Kandel, E. R., J. H. Schwartz, and T. M. Jessel. 2000. *Principles of Neuroscience*. 4th ed. New York: McGraw-Hill Medical.

Kawabata, H., and S. Zeki. 2004. Neural correlates of beauty. *Journal of Neurophysiology* 91:1699–1705.

Kimura, M. 1983. *The Neutral Theory of Molecular Evolution*. Cambridge, UK: Cambridge University Press.

King-Casas, B., D. Tomlin, C. Anen, C. F. Camerer, S. R. Quartz, and P. R. Montague. 2005. Getting to know you: Reputation and trust in a two-person economic exchange. *Science* 308:78–83.

Klyachko, V. A., and C. F. Stevens. 2003. Connectivity optimization and the positioning of cortical areas. *Proceedings of the National Academy of Sciences (USA)* 100:7937–7941.

Knill, D. C., and A. Pouget. 2004. The Bayesian brain: The role of uncertainty in neural coding and computation. *Trends in Neuroscience* 27(12):712–719.

Knuth, D. 2001. *The Art of Programming (Vols 1–3)*. Reading, Mass.: Addison-Wesley.

Knutson, B., A. Westdorp, E. Kaiser, and D. Hommer. 2000. FMRI visualization of brain activity during a monetary incentive delay task. *NeuroImage* 12:20–27.

Koch, C. 1999. *The Biophysics of Computation*. Oxford, UK: Oxford University Press.

Koch, C. 2004. *The Quest for Consciousness*. Englewood, Colo.: Roberts and Co.

Koch, C., and F. C. Crick. 2001. The zombie within. *Nature* 411:893.

Koch, C., T. Poggio, and V. Torre. 1982. Retinal ganglion cells: A functional interpretation of dendritic morphology. *Philosophical Transactions of the Royal Society of London B* 298:227–263.

Koch, C., T. Poggio, and V. Torre. 1983. Nonlinear interactions in a dendritic tree: Localization, timing, and role in information processing. *Proceedings of the National Academy of Sciences (USA)* 80(9):2799–2802.

Koepp, M. J., R. N. Gunn, A. D. Lawrence, V. J. Cunningham, and A. Dagher, et al. 1998. Evidence for striatal dopamine release during a video game. *Nature* 393:266–268.

Kosslyn, S. M., G. Ganis, and W. L. Thompson. 2003. Mental imagery: Against the nihilistic hypothesis. *Trends in Cognitive Science* 7(3):109–111.

Koulakov, A. A., and D. B. Chklovskii. 2001. Orientation preference maps in mammalian visual cortex: A wire length minimization approach. *Neuron* 29(2): 519–527.

Krebs, J. R., A. Kacelnick, and P. Taylor. 1978. Test of optimal sampling by foraging great tits. *Nature* 275:27–31.

Kringelbach, M. L., J. O'Doherty, E. T. Rolls, and C. Andrews. 2003. Activation of the human orbitofrontal cortex to a liquid food stimulus is correlated with its subjective pleasantness. *Cerebral Cortex* 13:1064–1071.

Kurzweil, R. 1999. *The Age of Spiritual Machines*. New York: Penguin.

Kurzweil, R. 2005. *The Singularity Is Near*. New York: Penguin.

Kwong, K. K., et al. 1992. Dynamic magnetic resonance imaging of human brain activity during primary sensory stimulation. *Proceedings of the National Academy of Sciences (USA)* 89:5675–5679.

Landauer, R. 1961. Irreversibility and heat generation in the computing process. *IBM Journal of Research and Development* 3:183–191.

Lane, R. D., E. M. Reiman, G. L. Ahern, G. E. Schwartz, and R. J. Davidson. 1997. Neuroanatomical correlates of happiness, sadness, and disgust. *American Journal of Psychiatry* 154(7):926–933.

Laughlin, S. B. 2001. Energy as a constraint on the coding and processing of sensory information. *Current Opinion Neurobiology* 11(4):475–480.

Laughlin, S. B., R. R. de Ruyter van Stevenick, and J. C. Anderson. 1998. The metabolic cost of neural information. *Nature Neuroscience* 1(1):36–41.

Laughlin, S. B., and T. J. Sejnowski. 2003. Communication in neuronal networks. *Science* 301:1870–1874.

Lavin, A., et al. 2005. Mesocortical dopamine neurons operate on distinct temporal domains using multimodal signaling. *Journal of Neuroscience* 25:5013–5023.

LeDoux, J. 1998. *The Emotional Brain*. New York: Simon and Schuster.

Leopold, D. A., and N. K. Logothetis. 1999. Multistable phenomena: changing views in perception. *Trends in Cognitive Science* 3(7):254–264.

Leopold, D. A., and A. Maier. 2006. Neuroimaging: perception at the brain's core. *Current Biology* 16(3):R95–98.

Leopold, D. A., A. Maier, and N. K. Logothetis. 2003. Measuring subjective visual perception in the nonhuman primate. *Journal of Consciousness Studies* 10(9–10): 115–130.

Leopold, D. A., A. J. O'Toole, T. Vetter, and V. Blanz. 2001. Prototype-referenced shape encoding revealed by high-level aftereffects. *Nature Neuroscience* 4(1): 89–94.

Leopold, D. A., and M. Wilke. 2005. Neuroimaging: Seeing the trees for the forest. *Current Biology* 15(18):R766–768.

Levy, W. B., and R. A. Baxter. 1996. Energy efficient neural codes. *Neural Computation* 8(3):531–543.

Levy, W. B., and R. A. Baxter. 2002. Energy-efficient neuronal computation via quantal synaptic failures. *Journal of Neuroscience* 22(11):4746–4755.

Li, Z., and J. J. Hopfield. 1989. Modeling the olfactory bulb and its neural oscillatory processings. *Biological Cybernetics* 61(5):379–392.

Ljungberg, T., P. Apicella, and W. Schultz. 1992. Responses of monkey dopamine neurons during learning of behavioral reactions. *Journal of Neurophysiology* 67:145–163.

Loewenstein, G. F., and R. H. Thaler. 1989. Anomalies: Intertemporal choice. *Journal of Economic Perspectives.* 3:181–93.

Logothetis, N. K. 1998. Single units and conscious vision. *Philosophical Transactions of the Royal Society of London B* 353:1801–1818.

Logothetis, N. K. 2000. Object recognition: Holistic representations in the monkey brain. *Spatial Vision* 13(2–3):165–178.

Logothetis, N. K. 2002. The neural basis of the blood-oxygen-level-dependent functional magnetic resonance imaging signal. *Philosophical Transactions of the Royal Society of London B* 357:1003–1037.

Logothetis, N. K. 2003. The underpinnings of the BOLD functional magnetic resonance imaging signal. *Journal of Neuroscience* 23(10):3963–3971.

Logothetis, N. K. 2004. Francis Crick, 1916–2004. *Nature Neuroscience* 7(10):1027–1028.

Logothetis, N. K., D. A. Leopold, and D. L. Sheinberg. 1996. What is rivaling during binocular rivalry? *Nature* 380:621–624.

Logothetis, N. K., J. Pauls, M. Augath, T. Trinath, and A. Oeltermann. 2001. Neurophysiological investigation of the basis of the fMRI signal. *Nature* 412:150–157.

Logothetis, N. K., and J. Pfeuffer. 2004. On the nature of the BOLD fMRI contrast mechanism. *Magnetic Resonance Imaging* 22(10):1517–1531.

Loomes, G., and R. Sugden. 1982. Regret theory: An alternative theory of rational choice under uncertainty. *The Economic Journal* 92:805–824.

Lopera, A., J. M. Buldu, M. C. Torrent, D. R. Chialvo, and J. García Ojalvo. 2006. Ghost stochastic resonance with distributed inputs in pulse-coupled electronic neurons. *Physical Review E* 73:021101.

Luce, R. D., and H. Raiffa. 1957. *Games and Decisions.* New York: Wiley.

MacKay, D.J.C. 2003. *Information Theory, Inference, and Learning Algorithms.* Cambridge, UK: Cambridge University Press.

Macrae, N. 1992. *John von Neumann.* New York: Pantheon Books.

Magee, J. C., and D. Johnston. 1997. A synaptically controlled, associative signal for Hebbian plasticity in hippocampal neurons. *Science* 275:209–213.

Magee, J. C., and D. Johnston. 2005. Plasticity of dendritic function. *Current Opinion in Neurobiology* 15(3):334–342.

Magill, R. 2003. *Motor Learning.* New York: McGraw-Hill.

Mainen, Z. F., J. Joerges, J. R. Huguenard, and T. J. Sejnowski. 1995. A model of spike initiation in neocortical pyramidal neurons. *Neuron* 15:1427–1439.

Mainen, Z. F., and T. J. Sejnowski. 1995. Reliability of spike timing in neocortical neurons. *Science* 268:1503–1506.

Maisels, C. K. 1990. *The Emergence of Civilization: From Hunting and Gathering to Agriculture.* London: Routledge.

Marr, D. 1969. A theory of cerebellar cortex. *Journal of Physiology* 202:437–470.

Marr, D. 1970. A theory for cerebral neocortex. *Proceedings of the Royal Society of London B* 176:161–234.

Marr, D. 1971. Simple memory: A theory for archicortex. *Philosophical Transactions of the Royal Society of London* 262:23–81.

Marr, D. 1975. Approaches to biological information processing. *Science* 190:875–876.

Marr, D., and T. Poggio. 1976. Cooperative computation of stereo disparity. *Science* 194:283–287.

Marr, D., and T. Poggio. 1979. A computational theory of human stereo vision. *Proceedings of the Royal Society of London B* 204:301–328.

Mauk, M., and D. Buonomono. 2004. The neural basis of temporal processing. *Annual Reviews of Neuroscience* 27:307–340.

Maunsell, J. H. 2004. Neuronal representations of cognitive state: reward or attention? *Trends in Cognitive Science* 8(6):261–265.

Mayer, E. 2001. *What Evolution Is.* New York: Basic Books.

McCabe, K., D. Houser, L. Ryan, V. Smith, and T. Trouard. 2001. A functional imaging study of cooperation in two-person reciprocal exchange. *Proceedings of the National Academy of Sciences (USA)* 98:11832–11835.

McCloskey, M. 1983. Intuitive physics. *Scientific American.* April:122–129.

McCloskey, M., A. Washburn, and L. Felch. 1983. Intuitive physics: the straight-down belief and its origin. *Journal of Experimental Psychology: Learning, Memory, and Cognition.* 9(4):636–649.

McClure, S. M., G. S. Berns, and P. R. Montague. 2003. Temporal prediction errors in a passive learning task activate human striatum. *Neuron* 38(2):339–346.

McClure, S. M., N. Daw, and P. R. Montague. 2003. A computational substrate for incentive salience. *Trends in Neuroscience* 26(8):423–428.

McClure, S. M., J. Li, D. Tomlin, K. S. Cypert, L. M. Montague, and P. R. Montague. 2004. Neural correlates of behavioral preference for culturally familiar drinks. *Neuron* 44:379–387.

Menzel, R., G. Leboulle, and D. Eisenhardt. 2006. Small brains, bright minds. *Cell* 124(2):237–239.

Miller, E. K., and J. D. Cohen. 2001. An integrative theory of prefrontal cortex function. *Annual Reviews of Neuroscience* 24:167–202.

Miller, P., and X. J. Wang. 2006. Power-law neuronal fluctuations in a recurrent network model of parametric working memory. *Journal of Neurophysiology* 95(2): 1099–1114.

Milo, R., S. Shen-Orr, S. Itzkovitz, N. Kashtan, D. Chklovskii, and U. Alon.Network. 2002. Network motifs: simple building blocks of complex networks. *Science* 298:824–827.

Mishkin, A. 2003. *Sojourner.* New York: Berkeley/Penguin Group.

Mitchison, G. 1991. Neuronal branching patterns and the economy of cortical wiring. *Proceedings of the Royal Society of London B* 245:151–158.

Montague, P. R. 1995. Integrating information at single synaptic connections. *Proceedings of the National Academy of Sciences (USA)* 92:2424–2425.

Montague, P. R., and G. S. Berns. 2002. Neural economics and the biological substrates of valuation. *Neuron* 36:265–284.

Montague, P. R., G. S. Berns, J. D. Cohen, S. M. McClure, G. Pagnoni, et al. 2002. Hyperscanning: Simultaneous fMRI during linked social interactions. *Neuroimage* 16:1159–1164.

Montague, P. R., P. Dayan, S. J. Nowlan, A. Pouget, and T. J. Sejnowski. 1993. Using aperiodic reinforcement for directed self-organization. *Advances in Neural Information Processing Systems* 5:969–976. San Mateo, Calif.: Morgan Kaufmann.

Montague, P. R., P. Dayan, C. Person, and T. J. Sejnowski. 1995. Bee foraging in an uncertain environment using predictive Hebbian learning. *Nature* 376:725–728.

Montague, P. R., P. Dayan, and T. J. Sejnowski. 1996. A framework for mesencephalic dopamine systems based on predictive Hebbian learning. *Journal of Neuroscience* 16:1936–1947.

Montague, P. R., S. E. Hyman, and J. D. Cohen. 2004. Computational roles for dopamine in behavioural control. *Nature* 431:760–767.

Montague, P. R., B. King-Casas, J. D. Cohen. 2006. Imaging valuation models in human choice. *Annual Reviews of Neuroscience* 29:417–448.

Montague, P. R., S. M. McClure, P. R. Baldwin, P. E. Phillips, E. A. Budygin, G. D. Stuber, M. R. Kilpatrick, and R. M. Wightman. 2004. Dynamic gain control of dopamine delivery in freely moving animals. *Journal of Neuroscience* 24(7):1754–1759.

Montague, P. R., and T. J. Sejnowski. 1994. The predictive brain: temporal coincidence and temporal order in synaptic learning mechanisms. *Learning Memory* 1:1–33.

Moravec, H. 1990. *Mind Children*. Cambridge: Harvard University Press.

Moravec, H. 1998. *Robot*. Oxford: Oxford University Press.

Nelson, P. 2003. *Biological Physics*. New York: W. H. Freeman and Co.

Niv, Y., N. D. Daw, P. Dayan. 2005. How hard to work: Response vigor, motivation and tonic dopamine. *Neural and Information Processing Systems* (in press).

Niv, Y., D. Joel, I. Meilijson, and E. Ruppin. 2002. Evolution of reinforcement learning in uncertain environments: A simple explanation for complex foraging behaviors. *Adaptive Behavior* 10(1):5–24.

Nowak, M., and K. Sigmund. 2005. Evolution of indirect reciprocity. *Nature* 437:1291–1298.

Nyquist, H. 1924. Certain factors affecting telegraph speed. *Bell System Technical Journal* 3:324–352.

Ochs, J., and A. E. Roth. 1989. An experimental study of sequential bargaining. *American Economic Review* 79:355–384.

O'Doherty, J. P., T. W. Buchanan, B. Seymour, R. J. Dolan. 2006. Predictive neural coding of reward preference involves dissociable responses in human ventral midbrain and ventral striatum. *Neuron* 49(1):157–166.

O'Doherty, J., P. Dayan, J. Schultz, R. Deichmann, K. Friston, and R. J. Dolan. 2004. Dissociable roles of ventral and dorsal striatum in instrumental conditioning. *Science* 304:452–454.

Ogawa, S., T. M. Lee, A. R. Kay, and D. W. Tank. 1990b. Brain magnetic resonance imaging with contrast dependent on blood oxygenation. *Proceedings of the National Academy of Sciences (USA)* 87:9868–9872.

Ogawa, S., T. M. Lee, A. S. Nayak, and P. Glynn. 1990a. Oxygenation-sensitive contrast in magnetic resonance image of rodent brain at high magnetic fields. *Magnetic Resonance in Medicine* 14:68–78.

Ogawa, S., R. S. Menon, D. W. Tank, S. G. Kim, H. Merkle, et al. 1993. Functional brain mapping by blood oxygenation level–dependent contrast magnetic resonance imaging. *Biophysics Journal* 64:803–812.

Ogawa, S., D. W. Tank, R. Menon, J. M. Ellermann, S. G. Kim, et al. 1992. Intrinsic signal changes accompanying sensory stimulation: Functional brain mapping with magnetic resonance imaging. *Proceedings of the National Academy of Sciences (USA)* 89:5951–5955.

Ohyama, T., W. L. Nores, M. Murphy, and M. D. Mauk. 2003. What the cerebellum computes. *Trends in Neuroscience* 26(4):222–227.

Olds, J. 1958. Self-stimulation of the brain. *Science* 127:315–324.

Olds, J., and P. Milner. 1954. Positive reinforcement produced by electrical stimulation of septal area and other regions of rat brain. *Journal of Comparative Physiology and Psychology* 47:419–427.

Olshausen, B. A., and D. J. Field. 1996. Emergence of simple-cell receptive field properties by learning a sparse code for natural images. *Nature* 381:607–609.

Olshausen, B. A., and D. J. Field. 2004. Sparse coding of sensory inputs. *Current Opinion in Neurobiology* 14(4):481–487.

Olshausen, B. A., and D. J. Field. 2005. How close are we to understanding v1? *Neural Computation* 17(8):1665–1699.

O'Reilly, R. C., T. S. Braver, and J. D. Cohen. 1999. A biologically-based neural network model of working memory. In *Models of Working Memory: Mechanisms of Active Maintenance and Executive Control* (eds. A. Miyake and P. Shah) New York: Cambridge University Press, pp. 375–411.

O'Reilly, R. C., D. C. Noelle, T. S. Braver, and J. D. Cohen. 2002. Prefrontal cortex and dynamic categorization tasks: Representational organization and neuromodulatory control. *Cerebral Cortex* 12:246–257.

Otani, S., ed. 2004. *The Prefrontal Cortex.* Norwell, Mass.: Kluwer Academic Publishers.

Ozdenoren, E., S. W. Salant, and D. Silverman. May 31, 2005. Willpower and the optimal control of visceral urges. http://ssrn.com/abstract=742564

Packard, M. G., and B. J. Knowlton. 2002. Learning and memory functions of the basal ganglia. *Annual Reviews of Neuroscience* 25:563–593.

Pauling, L., and M. Delbruck. 1940. The nature of intermolecular forces operative in biological processes. *Science* 92:77–79.

Penrose, R. 1989. *The Emperor's New Mind.* Oxford: Oxford University Press.

Penrose, R. 1994. *Shadows of the Mind.* Oxford: Oxford University Press.

Perkins, R. 1997. *Cosmic Suicide*. Dallas: Pentaradial Press.

Phillips, P.E.M., G. D. Stuber, M.L.A.V. Heien, R. M. Wightman, and R. M. Carelli. 2003. Subsecond dopamine release promotes cocaine seeking. *Nature* 422: 614–616.

Pinker, S. 1997. *How the Mind Works*. New York: W. W. Norton & Co.

Pinker, S. 2002. *The Blank Slate*. New York: Penguin.

Platt, M. L., and P. W. Glimcher. 1999. Neural correlates of decision variables in parietal cortex. *Nature* 400:233–238.

Poggio, T., V. Torre, and C. Koch. 1985. Computational vision and regularization theory. *Nature* 317:314–319.

Potenza, M. N., et al. 2003. Gambling urges in pathological gambling: a functional magnetic resonance imaging study. *Archives of General Psychiatry* 60:828–836.

Pouget, A., P. Dayan, R. S. Zemel. 2003. Inference and computation with population codes. *Annual Reviews of Neuroscience* 26:381–410.

Prelec, D. 2004. A Bayesian truth serum for subjective data. *Science* 306:426–462.

Quartz, S. R., P. Dayan, P. R. Montague, and T. J. Sejnowski. 1992. Expectation learning in the brain using diffuse ascending connections. *Society for Neuroscience Abstracts* 18:1210.

Quartz, S. R., and T. J. Sejnowski. 2002. *Liars, Lovers, Heroes*. New York: HarperCollins.

Quian Quiroga, R., L. Reddy, G. Kreiman, C. Koch, and I. Fried. 2005. Invariant visual representation by single neurons in the human brain. *Nature* 435:1102–1107.

Rachlin, H. 2000. *The Science of Self-Control*. Cambridge, Mass.: Harvard University Press.

Raichle, M. E. 2003. Functional brain imaging and human brain function. *Journal of Neuroscience* 23(10):3959–3962.

Raichle, M. E., and D. A. Gusnard. 2002. Appraising the brain's energy budget. *Proceedings of the National Academy of Sciences (USA)* 99(16):10237–10239.

Ramachandran, V. S. 2004. *A Brief Tour of Human Consciousness*. New York: Pi Press.

Ramachandran, V. S., and S. Blakeslee. 1998. *Phantoms in the Brain*. New York: William Morrow.

Ramachandran, V. S., and E. M. Hubbard. 2003. Hearing colors, tasting shapes. *Scientific American* 288(5): 52–59.

Ramón y Cajal, S. 1995. *Histology of the Nervous System of Man and Vertebrates*, trans. L. Azoulay, N. Swanson, and L. Swanson. New York: Oxford University Press.

Rapoport, A., and A. M. Chammah. 1965. *Prisoner's Dilemma: A Study in Conflict and Cooperation*. Ann Arbor: University of Michigan Press.

Redish, D. 2004. Addiction as a computational process gone awry. *Science* 306: 1944–1947.

Reill, P. H. 2005. *Vitalizing Nature in the Enlightenment*. Berkeley: University of California Press.

Ridley, M. 1997. *The Origins of Virtue*. New York: Viking Press.

Rieke, F., D. Warland, R. de Ruyter van Steveninck, and W. Bialek. 1997. *Spikes: Exploring the Neural Code*. Cambridge, Mass.: MIT Press.

Rilling, J. K., D. A. Gutman, T. R. Zeh, G. Pagnoni, G. S. Berns, and C. D. Kilts. 2002. A neural basis for social cooperation. *Neuron* 35:395–405.

Rilling, J. K., A. G. Sanfey, J. A. Aronson, L. E. Nystrom, and J. D. Cohen. 2003. The neural basis of economic decision-making in the ultimatum game. *Science* 300: 1755–1758.

Rilling, J. K., A. G. Sanfey, J. A. Aronson, L. E. Nystrom, and J. D. Cohen. 2004. The neural correlates of theory of mind within interpersonal interactions. *Neuroimage* 22:1694–1703.

Roberts, A. C., T. W. Robbins, and L. Weiskrantz. 1998. *The Prefrontal Cortex*. Oxford: Oxford University Press.

Robinson, T., and K. Berridge. 1993. The neural basis of drug craving: An incentive-sensitization theory of addiction. *Brain Research Reviews* 18(3):247–291.

Robinson, T., and K. Berridge. 2003. Addiction. *Annual Reviews of Psychology* 54: 25–53.

Roese, N. J. 2005. *If Only*. New York: Broadway Books.

Roese, N. J., and A. Summerville. 2005. What we regret most . . . and why. *Personality and Social Psychology Bulletin*. 31:1273–1285.

Rolls, E. T. 2000. *The Brain and Emotion*. Oxford: Oxford University Press.

Rolls, E. T. 2000. Memory systems in the brain. *Annual Reviews of Psychology* 51:599–630.

Rolls, E. T. 2000. The orbitofrontal cortex and reward. *Cerebral Cortex* 10:284–294.

Romo, R., and W. Schultz. 1990. Dopamine neurons of the monkey midbrain: contingencies of responses to active touch during self-initiated arm movements. *J Neurophysiology* 63(3):592–606.

Rorie, A. E., and W. T. Newsome. 2005. A general mechanism for decision-making in the human brain? *Trends in Cognitive Science* 9:41–43.

Rosenstein, M. T., and A. G. Barto. 2004. Supervised actor-critic reinforcement learning. In *Learning and Approximate Dynamic Programming: Scaling up to the Real World*, ed. J. Si, A. Barto, W. Powell, D. Wunsch, pp. 359–80. New York: Wiley.

Roth, A. 1995. Bargaining experiments. In *Handbook of Experimental Economics*, ed. J. H. Kagel, A. E. Roth, pp. 253–348. Princeton: Princeton University Press.

Rougier, N. P., and R. C. O'Reilly. 2002. Learning representations in a gated prefrontal cortex model of dynamic task switching. *Trends in Cognitive Science* 26:503–520.

Rumelhart, D. E., and J. McClelland. 1986. *Parallel Distributed Processing: Explorations in the Microstructure of Cognition*. Cambridge, Mass.: MIT Press.

Salinas, E., and L. F. Abbott. 2001. Coordinate transformations in the visual system: how to generate gain fields and what to compute with them. *Progress in Brain Research* 130:175–190.

Sanfey, A. G., J. K. Rilling, J. A. Aronson, L. E. Nystrom, and J. D. Cohen. 2003. The neural basis of economic decision making in the ultimatum game. *Science* 300:1755–1757.

Sangrey, T. D., W. O. Friesen, and W. B. Levy. 2004. Analysis of the optimal channel density of the squid giant axon using a reparameterized Hodgkin-Huxley model. *Journal of Neurophysiology* 91(6):2541–2550.

Scannell, J. W., C. Blakemore, and M. P. Young. 1995. Analysis of connectivity in the cat cerebral cortex. *Journal of Neuroscience* 15(2):1463–1483.

Scannell, J. W., and M. P. Young. 1993. The connectional organization of neural systems in the cat cerebral cortex. *Current Biology* 3(4):191–200.

Schaal, S., and N. Schweighofer. 2005. Computational motor control in humans and robots. *Current Opinion in Neurobiology* 15(6):675–682.

Schneidman, E., W. Bialek, and M. J. Berry. 2003. Synergy, redundancy, and independence in population codes. *Journal of Neuroscience* 23(37):11539–11553.

Schultz, W., P. Apicella, and T. Ljungberg. 1993. Responses of monkey dopamine neurons to reward and conditioned stimuli during successive steps of learning a delayed response task. *Journal of Neuroscience* 13(3):900–913.

Schultz, W., P. Dayan, and P. R. Montague. 1997. A neural substrate of prediction and reward. *Science* 275:1593–1599.

Schultz, W., and A. Dickinson. 2000. Neuronal coding of prediction errors. *Annual Reviews of Neuroscience* 23:473–500.

Schultz, W., and R. Romo. 1990. Dopamine neurons of the monkey midbrain: contingencies of responses to stimuli eliciting immediate behavioral reactions. *Journal of Neurophysiology* 63(3):607–624.

Searle, J. 1980. Minds, brains, and programs. *Behavioral and Brain Sciences* 3:417–424.

Searle, J. 1984. *Minds, Brains, and Science.* Cambridge, Mass.: Harvard University Press.

Sejnowski, T. J. 1976a. On global properties of neuronal interaction. *Biological Cybernetics* 22:85–95.

Sejnowski, T. J. 1976b. On the stochastic dynamics of neuronal interaction. *Biological Cybernetics* 22:203–211.

Sejnowski, T. J. 1977a. Storing covariance with nonlinearly interacting neurons. *Journal of Mathematical Biology* 4:303–321.

Sejnowski, T. J. 1977b. Statistical constraints on synaptic plasticity. *Journal of Theoretical Biology.* 69:385–389.

Sejnowski, T. J., P. K. Kienker, and G. E. Hinton. 1986. Learning symmetry groups with hidden units: Beyond the perception. *Physica D* 22:260–275.

Sejnowski, T. J., C. Koch, and P. S. Churchland. 1988. Computational neuroscience. *Science.* 241:1299–1306.

Seymour, B., J. P. O'Doherty, P. Dayan, M. Koltzenburg, A. K. Jones, R. J. Dolan, K. J. Friston, and R. S. Frackowiak. 2004. Temporal difference models describe higher-order learning in humans. *Nature* 429:664–667.

Shadlen, M. N., and W. T. Newsome. 2001. Neural basis of a perceptual decision in the parietal cortex (area LIP) of the rhesus monkey. *Journal of Neurophysiology* 86:1916–1936.

Shallice, T., and P. Burgess. 1991. Higher-order cognitive impairments and frontal lobe lesions in man. In *Frontal Lobe Function and Dysfunction*, ed. H. S. Levin, H. M. Eisenberg, A. L. Benton, pp. 125–38. New York: Oxford University Press.

Shannon, C. E. 1940. An algebra for theoretical genetics. Ph.D. dissertation, MIT.

Shannon, C. E. 1948. A mathematical theory of communication. *Bell System Technical Journal* 27:379–423, 623–656.

Shear, J., ed. 1997. *Explaining Consciousness*. Cambridge, Mass.: MIT Press.

Shefrin, H. M., and R. H. Thaler. 1988. The behavioral life-cycle hypothesis. *Economic Inquiry* 26:609–643.

Shepard, R. N. 1990. *Mind Sights*. New York: W. H. Freeman.

Shepard, R. N. 2001. Perceptual-cognitive universals as reflections of the world. *Behavioral Brain Sciences* 24:581–601.

Shepard, R. N., and L. A. Cooper. 1982. *Mental Images and Their Transformations*. Cambridge: MIT Press.

Shizgal, P. 1997. Neural basis of utility estimation. *Current Opinion in Neurobiology* 7:198–208.

Sigmund, K., E. Fehr, and M. A. Nowak. 2002. The economics of fair play. *Scientific American* 286:82–87.

Simon, H. 1955. A behavioral model of rational choice. *Quarterly Journal of Economics* 69:99–118.

Simoncelli, E. P., and B. A. Olshausen. 2001. Natural image statistics and neural representation. *Annual Reviews of Neuroscience* 24:1193–1216.

Singer, T., B. Seymour, J. P. O'Doherty, K. E. Stephan, R. J. Dolan, and C. D. Frith. 2006. Empathic neural responses are modulated by the perceived fairness of others. *Nature* 439:466–469.

Sipser, M. 2005. *Introduction to the Theory of Computation*, 2nd Ed. Thomson Course Technology.

Smiley, G. 2003. *Rethinking the Great Depression*. Chicago: Ivan R. Dee.

Smith, A. 1991. *The Wealth of Nations*. Amherst, NY.: Prometheus Books.

Smith, A. J., H. Blumenfeld, K. L. Behar, D. L. Rothman, R. G. Shulman, and F. Hyder. 2002. Cerebral energetics and spiking frequency: the neurophysiological basis of fMRI. *Proceedings of the National Academy of Sciences (USA)* 99 (16):10765–10770.

Softky, W. R., and C. Koch. 1993. The highly irregular firing of cortical cells is inconsistent with temporal integration of random EPSPs. *Journal of Neuroscience* 13:334–350.

Sporns, O., D. R. Chialvo, M. Kaiser, and C. C. Hilgetag. 2004. Organization, development and function of complex brain networks. *Trends in Cognitive Science* 8(9):418–425.

Sporns, O., and R. Kötter. 2004. Motifs in brain networks. *PLoS Biology* 2(11):e369.

Sporns, O., G. Tononi, and R. Kotter. 2005. The human connectome: a structural description of the human brain. *PLoS Computational Biology* 1(4):e42.

Sporns, O., and J. D. Zwi. 2004. The small world of the cerebral cortex. *Neuroinformatics* 2(2):145–162.

Stuss, D. T., and R. Knight. 2002. *Principles of Frontal Lobe Function*. Oxford: Oxford University Press.

Sugrue, L. P., G. S. Corrado, and W. T. Newsome. 2004. Matching behavior and the representation of value in the parietal cortex. *Science* 304:1782–1787.

Sugrue, L. P., G. S. Corrado, and W. T. Newsome. 2005. Choosing the greater of two goods: neural currencies for valuation and decision making. *Nature Reviews Neuroscience* 6:1–13.

Sutton, R. S. 1998. Learning to predict by the methods of temporal difference. *Machine Learning* 3:9–44.

Sutton, R. S., and A. G. Barto. 1998. *Reinforcement Learning*. Cambridge, Mass.: MIT Press.

Tank, D. W., and J. J. Hopfield. 1987. Collective computation in neuronlike circuits. *Scientific American* 257(6):104–114.

Tavel, M. A. 1971. Invariant variation problems. *Transport Theory and Statistical Mechanics* 1(3):183–207. Translation of Emmy Noether's paper of 1918, "Invariante Variationsprobleme," Nachr. v. d. Ges. d. Wiss. zu Göttingen, pp. 235–257.

Tedrake, R. L. 2004. Applied optimal control for dynamically stable legged locomotion. Ph.D. thesis, Massachusetts Institute of Technology.

Tedrake, R., T. W. Zhang, and H. S. Seung. 2004. Stochastic policy gradient reinforcement learning on a simple 3D biped. In *Proceedings of the IEEE International Conference on Intelligent Robots and Systems (IROS)*, p. 2849–2854, Senda, Japan.

Tesauro, G. 1989. Neurogammon wins Computer Olympiad. *Neural Computation* 1:321–323.

Tesauro, G. 1995. Temporal difference learning and TD-Gammon. *Communications of the ACM* 38(3):58–68.

Tesauro, G., and T. J. Sejnowski. 1989. A parallel network that learns to play backgammon. *Artificial Intelligence* 39:357–390.

Tolias, A. S., T. Moore, S. M. Smirnakis, E. J. Tehovnik, A. G. Siapas, and P. H. Schiller. 2001. Eye movements modulate visual receptive fields of V4 neurons. *Neuron* 29(3):757–767.

Tolias, A. S., S. M. Smirnakis, M. A. Augath, T. Trinath, and N. K. Logothetis. 2001. Motion processing in the macaque: revisited with functional magnetic resonance imaging. *Journal of Neuroscience* 21(21):8594–8601.

Tononi, G. 2005. Consciousness, information integration, and the brain. *Progress in Brain Research* 150:109–126.

Trivers, R. 2002. *Natural Selection and Social Theory*. Oxford: Oxford University Press.

Trivers, R. 2004. Mutual benefits at all levels of life. *Science* 304:964–965.

Turing, A. M. 1936. On computable numbers, with an application to the Entscheidungsproblem. *Proceedings of the London Mathematical Society* 2(42):173–198.

Turing, A. M. 1939. Systems of logic based on ordinals. *Proceedings of the London Mathematical Society* 45:161–228.

Tversky, A., and D. Kahneman. 1974. Judgment under uncertainty: heuristics and biases. *Science* 185:1124–1131.

Tversky, A., and D. Kahneman. 1981. The framing of decisions and the psychology of choice. *Science* 211:453–458.

Usher, M., J. D. Cohen, D. Servan-Schreiber, J. Rajkowski, and G. Aston-Jones. 1999. The role of locus coeruleus in the regulation of cognitive performance. *Science* 283:549–554.

Van Essen, D. C. 1996. Formation and folding of cerebral and cerebellar cortex: A tension-based theory of morphogenesis and compact wiring. *Society for Neuroscience Abstracts* 22:1013.

Van Essen, D.C. 1997. A tension-based theory of morphogenesis and compact wiring in the central nervous system. *Nature* 385:313–318.

Vartanian, O., and V. Goel. 2004. Neuroanatomical correlates of aesthetic preference for paintings. *Neuroreport* 15:893–897.

Von Neumann, J. 1932. *Mathematical Foundations of Quantum Mechanics*. Trans. R. T. Beyer, Princeton: Princeton University Press.

Von Neumann, J. (edited by A.W. Burks). 1966. *Theory of Self-Reproducing Automata*. Urbana: University of Illinois Press.

Von Neumann, J., and O. Morgenstern. 1944. *The Theory of Games and Economic Behavior*, Princeton: Princeton University Press.

Waelti, P., A. Dickinson, and W. Schultz. 2001. Dopamine responses comply with basic assumptions of formal learning theory. *Nature* 412:43–48.

Watkins, C.J.C.H. 1989. *Learning from delayed rewards*. Ph.D. thesis, King's College, Cambridge University.

Watkins, C.J.C.H., and P. Dayan. 1992. Q-learning. *Machine Learning* 8:279–292.

Weiss, Y., E. P. Simoncelli, and E. H. Adelson. 2002. Motion illusions as optimal percepts. *Nature Neuroscience* 5(6):598–604.

Weng, J., J. McClelland, A. Pentland, O. Sporns, I. Stockman, M. Sur, and E. Thelen. 2001. Artificial intelligence. Autonomous mental development by robots and animals. *Science* 291:599–600.

Wightman, R. M. 2006. Detection technologies. Probing cellular chemistry in biological systems with microelectrodes. *Science* 311:1570–1574.

Wilkinson, G. S. 1984. Reciprocal food sharing in the vampire bat. *Nature* 308:181–184.

Wilkinson, G. S. 1990. Food sharing in vampire bats. *Scientific American*. February:76–82.

Williams, G. C. 1966. *Adaptation and Natural Selection*. Princeton: Princeton University Press.

Wilson, E. O. 1975; 2000. *Sociobiology*. Cambridge, Mass.: Harvard University Press.

Wilson, E. O. 1978. *On Human Nature*. Cambridge, Mass.: Harvard University Press.

Wilson, E. O. 1998. *Consilience*. New York: Vintage/Random House.

Wise, R. A., and M. A. Bozarth. 1987. A psychomotor stimulant theory of addiction. *Psychological Review* 94:469–492.

Wolfe, T. 1996. Sorry but your soul just died. *Forbes* magazine.

Wolfram, S. 2002. *A New Kind of Science*. Champaign, Ill.: Wolfram Media Inc.

Wolpert, D., and Z. Ghahramani. 2000. Computational principles of movement neuroscience. *Nature Neuroscience* 3:1212–1217.

Wong, K. F., and X. J. Wang. 2006. A recurrent network mechanism of time integration in perceptual decisions. *Journal of Neuroscience* 26(4):1314–1328.

Yellot, J. I. 1983. Spectral consequences of photoreceptor sampling in the rhesus retina. *Science* 221:382–385.

Young, M. P. 1992. Objective analysis of the topological organization of the primate cortical visual system. *Nature* 358:152–155.

Young, M. P., J. W. Scannell, M. A. O'Neill, C. C. Hilgetag, G. Burns, and C. Blakemore. 1995. Non-metric multidimensional scaling in the analysis of neuroanatomical connection data and the organization of the primate cortical visual system. *Philosophical Transactions of the Royal Society of London B* 348: 281–308.

Yu, A. J., and P. Dayan. 2005. Uncertainty, neuromodulation, and attention. *Neuron* 46:681–692.

Zeki, S., J. D. Watson, C. J. Lueck, K. J. Friston, C. Kennard, and R. S. Frackowiak. 1991. A direct demonstration of functional specialization in human visual cortex. *Journal of Neuroscience* 11(3):641–649.

INDEX